Emerging Technologies in Clean Energy 2018

Topical Conference at the 2018 AIChE Spring Meeting and
14th Global Congress on Process Safety

Orlando, Florida, USA
22 – 25 April 2018

ISBN: 978-1-5108-6431-3

Emerging Technologies in Clean Energy 2018

Topical Conference at the 2018 AIChE Spring Meeting and
14th Global Congress on Process Safety

Orlando, Florida, USA
22 – 25 April 2018

AIChE
The Global Home of Chemical Engineers

Printed from e-media with permission by:

Curran Associates, Inc.
57 Morehouse Lane
Red Hook, NY 12571

Some format issues inherent in the e-media version may also appear in this print version.

Printed by Curran Associates, Inc. (2018)

For permission requests, please contact AIChE
at the address below.

AIChE
120 Wall Street, FL 23
New York, NY 10005-4020

Phone: (800) 242-4363
Fax: (203) 775-5177

www.aiche.org

Additional copies of this publication are available from:

Curran Associates, Inc.
57 Morehouse Lane
Red Hook, NY 12571 USA
Phone: 845-758-0400
Fax: 845-758-2633
Email: curran@proceedings.com
Web: www.proceedings.com

TABLE OF CONTENTS

Some abstracts in this book refer to images that were not made available by the publisher

513973 Review on Vanadium Redox Flow Battery Technologies: Current Trends and Future Directions

Abstract:

Renewable energy technologies are becoming increasingly important to help mitigate the environmental impacts of energy production and to preserve the planet for future generations. In accordance with these objectives, several electrochemical energy storage technologies, including lithium-ion, sodium-sulfur and ZEBRA batteries have been proposed and deployed in field applications. For all these rechargeable batteries, coupled energy storage capacity and power rating is a common drawback limiting their widespread implementation. In contrast, redox flow batteries offer the advantage of decoupled power and energy ratings, and the ability to adjust them for specific operational requirements (especially for energy centric applications).

During the last 30 years all-vanadium redox flow batteries (VRFB) have evolved from pioneering academic research to optimized and reliable energy storage systems with high cycle stability and longevity. Furthermore, VRFB technology is intrinsically safe, and the risk of thermal runaway is readily avoided. VRFBs are designed to effectively store and release energy for more than 20 years. For a comprehensive analysis of the total cost of ownership, it is necessary to consider long-term operating costs and system efficiency since they will become more significant over the lifecycle of the system than initial capital investment for acquisition and installation. Therefore, it is critical to optimize all the components of the battery to enhance its efficiency, durability and cycle stability.

This review starts with discussing basic thermodynamic terms, which are voltage- (VE), charge- (CE) and energy efficiency (EE). The voltage efficiency is correlated with the internal resistance of a stack or single cell. On this basis, the influence of each component in the electrochemical energy converter is discussed. The overall resistance can be separated into fractions with electronic and ionic contributions arising from system components such as, current collector, bipolar plate, electrode, membrane and electrolyte. Additionally, the contributions of the electrochemistry (reaction overvoltage) and mass transport have to be considered, which affect the voltage efficiency as well. To minimize the electrochemical overvoltage both the carbon-based electrode material and the electrolyte can be optimized, which is accomplished by methods of activation and new electrolyte mixtures, respectively, resulting in reduced charge transfer resistance for the vanadium reactions. The recent state of the art is analyzed by a literature review, and new developing technologies are discussed. The majority of VRFB demonstrations have been based on the cells with carbon felt electrodes, which provide excellent conditions for electrolyte flow. However, in recent years, carbon papers are introduced as an alternative to carbon felt electrodes, which are thin with a high surface area. In this review, the benefits and shortcomings of both approaches are discussed. It should also be noted that regardless of electrode type almost all VRFB systems reported in the literature require a membrane to separate positive and negative electrolyte. However, membranes often contribute to total resistance with the highest percentage. To overcome this drawback, new types of electrolytes (immiscible) have been discussed in the recent literature which enable membrane-free flow cell designs. Overall, the objective of this review is to offer a summary of recent developments and best design practices in the field of VRFBs.

513974 Optical Techniques to Illuminate Improvements in Aqueous Organic Redox Flow Batteries

Abstract:

The global increase in renewable energy generation penetration has spurred research in emerging redox flow batteries (RFBs) technologies to address variability in electricity supply. Novel chemistries for aqueous organic RFBs are being explored as potentially inexpensive, long-lifetime electrolytes. However, both electrolyte and hardware engineering challenges remain for RFBs, which limit their feasibility for grid-scale operation. In this study we use optical techniques including absorption spectroscopy and fluorescence microscopy, coupled with electrochemistry, to illuminate molecular degradation, active species crossover, intermolecular interactions, and heterogeneous mass transport within operating RFBs. Through each of these, we develop electrolyte and hardware engineering solutions to significantly improve aqueous organic RFB performance. These optical techniques provide unique insight that is applicable to a variety of RFB chemistries and architectures, pushing the envelope toward a clean energy future.

513047 Commercial Perfluorosulfonic Acid Membranes for Vanadium Redox Flow Battery

Abstract:

Vanadium redox flow batteries (VRFB) are nowadays successfully integrated as energy storage onto grid networks in dozens of demonstrational projects. High efficiency (80% DC-DC), versatile arrangement of decoupled power (kW) to capacity (kWh), extended durability and fast demand response predetermine VRFB as a great, cheap and competitive stationary energy storage solution allowing compensation of hardly predictable non-stabilities in output power of photovoltaics or wind turbines on the power transmission level. However, the broader commercialization of the technology is still obstructed by relatively high investment costs.

Ion-exchange membrane is a key component of VRFB cell as it directly influences the power and efficiency of the stack â€" the resistance of the typically used cation-exchange membrane counts for up to 50 % of total internal resistance of the battery stack. At the same time, price of the membrane may exceed 25 % of the overall costs. Thus, the identification of highly conductive, low-cost membrane with minimal permeability for vanadium ions and sufficient durability in acidic electrolytes is vitally needed for VRFB developers.

In our contribution we present the results of a broad systematic study focused on the effect of membrane properties (charge, ion-exchange capacity (IEC), thickness) on VRFB operation. The various types of commercially available ion-exchange membranes were characterized with the respect to the properties relevant for VRFB operation such as: i) through-plane ionic conductivity in the environment of VRFB cell, ii) permeability for vanadium ions of different oxidation states and iii) performance in VRFB single-cell.

Results obtained from permeation measurement showed that the permeation of V^{2+} ions contributes most to the self-discharge losses of VRFB from all vanadium cations. According to our assumptions, the conductivity and selectivity of membrane significantly depend on the concentration of ion-exchange groups in the membrane. However, these properties are also influenced by the specific inner arrangement of hydrophobic and hydrophilic domains in the material. Small- and wide-angle X-ray scattering of fully hydrated membranes revealed the effect of polymer molecular architecture on the size of hydrophilic domains. The size of hydrophilic domains is a consequence of different side-chain lengths in tested polymers.

The observed trends were also confirmed by the characterization of membranes in VRFB single-cell. It was found that optimum membrane properties depend strongly on the VRFB target application. Energy efficiency evaluation showed that membranes with lower IEC are suitable for batteries operating at lower current densities when reduction of vanadium cross-over is important. Oppositely, membranes with higher IEC are more suitable for VRFB operating at higher current densities. Membrane inner morphology must be always considered. For some VRFB applications, the suppressed operating costs (related to electrolytes rebalancing) with less permeable membrane can, in the long-term perspective, compensate higher investment costs due to lower power densities at given efficiency.

513128 Role of Hydrogen Evolution Reaction during Zinc Electrodeposition from Flowing Alkaline Zincate Solutions

Abstract:

Zinc electrodes in alkaline solutions have been tested over time in a combination with various redox couples in primary and secondary cells. A very promising use of a zinc seems to be in flow batteries, where a flow of an electrolyte can solve a dendritic growth of the zinc metal, which is a main obstacle that prevents a successful development of the secondary batteries based on the alkaline zinc chemistry. Besides that, an attention has to be paid to a presence of hydrogen evolution reaction, which is known as a parasitic reaction during the zinc electrodeposition from an alkaline zincate solutions. The hydrogen evolution reaction is responsible for current efficiency losses during the electrolysis, but rising hydrogen bubbles also cause an extra convection within a diffusion layer, which leads to an enhanced mass transport of zincate ions to an electrode surface.

This work is aimed to determine an influence of the hydrogen evolution reaction on the zinc electrodeposition from the flowing alkaline zincate solutions. The study consisted of experimental electrodepositions of the zinc metal in an electrolyzer with the convective flow of the electrolyte and a subsequent data evaluation by mathematical models. The results show a complex role of the hydrogen evolution reaction during the zinc electrodeposition from the flowing alkaline zincate solutions, which complicates usage of a limiting current density as a morphology indicator of the zinc electrodeposition.

513455 Graphite Felt Electrodes for Vanadium Redox Flow Battery: Long-Term Stability and Hydrogen Evolution Reaction

Abstract:

Vanadium redox flow battery (VRFB) represents a promising electrochemical energy storage solution for various applications and scales ranging from few-kW residential batteries to hundreds-MW grid stabilization sites. For further commercialization of the technology, the intensification of the reactions on porous electrodes is required. Intensification is often achieved by thermal treatment or catalyst deposition. Long-term stability of catalytic activity of thermally treated felt is very important for desired battery life time of more than 10 years. In this contribution, we carefully analyze the effect of the electrode potential and battery cycling on felt stability. Various felt types were characterized by a complex set of physico- (BET, XPS, Raman, SEM) and electro-chemical (CV, EIS) techniques and the effect of felt degradation on the battery performance and hydrogen evolution reaction was systematically evaluated. Labscale single-cell with inner platinum sense electrodes was modified to specifically test the felt at studied electrode potentials and also to characterize the performance of positive and negative electrode separately. The labscale battery system was modified to allow quantification of hydrogen evolution side reaction. The slow continual deactivation of negative electrode under long-term battery operation was found to be dependent on active surface area.

Abstract:

Negative Electrolyte Additive and Heteroatom Co-Doping of Positive Electrode for Improving Electrical Performance of Vanadium Redox Flow Battery

Abstract

Photovoltaics, wind and nuclear power are the sources for satisfying the energy need of the growing economy. The renewable energy technologies feeds significant electricity to the grid intermittently and hence effective energy storage technologies are required for load shifting. Redox flow batteries (RFBs) is an energy storage technology revisited for the above said purpose. Among RFBs, vanadium redox flow battery (VRFB) is the most matured in terms of development and deployment.

However, it is suffering with issues like low capacity retention ADDIN EN.CITE ADDIN EN.CITE.DATA

3C456E644E6F74653E3C436974653E3C417574686F723E4A69613C2F417574686F723E3C596561723E323031343C2F596561723E3C526563
[1]

3C456E644E6F74653E3C436974653E3C417574686F723E4A69613C2F417574686F723E3C596561723E323031343C2F596561723E3C526563

and poor round-trip energy efficiency (EE). EE is a product of coulombic and voltage efficiency (VE). To address these issues, researchers have tried various additives to the positive electrolyte ADDIN EN.CITE <EndNote><Cite><Author>Corcuera</Author><Year>2012</Year><RecNum>4</RecNum><DisplayText><style and electrolyte rebalancing methods for the vanadium redox flow battery</title><secondary-title>European Chemical Bulletin</secondary-title></titles><periodical><full-title>European Chemical Bulletin</full-title></periodical><pages>511-519</pages><volume>1</volume><number>12</number><dates><year>2012</year></dates><urls></urls></record></Cite></EndNote>[2] and electrode activation strategies ADDIN EN.CITE <EndNote><Cite><Author>Wu</Author><Year>2016</Year><RecNum>5</RecNum><DisplayText><style db-id="t55rvpwf85zw9weptavpaxzsradpxfv0vsew">5</key></foreign-keys><ref-type name="Journal Article">17</ref-type><contributors><authors><author>Wu, Lantao</author><author>Shen, Yi</author><author>Yu, Lihong</author><author>Xi, Jingyu</author><author>Qiu, Xinping</author></authors></contributors><titles><title>Boosting vanadium flow battery performance by nitrogen-doped carbon nanospheres electrocatalyst</title><secondary-title>Nano Energy</secondary-title></titles><periodical><full-title>Nano Energy</full-title></periodical><pages>19-28</pages><volume>28</volume><dates><year>2016</year></dates><isbn>2211-2855</isbn><urls></urls></record></Cite><Cite><Author>Kabtamu</Author><Year>2016</Year><RecNum>3</RecNum><record><rec-number>3</rec-number><foreign-keys><key app="EN" db-id="t55rvpwf85zw9weptavpaxzsradpxfv0vsew">3</key></foreign-keys><ref-type name="Journal Article">17</ref-type><contributors><authors><author>Kabtamu, Daniel Manaye</author><author>Chen, Jian-Yu</author><author>Chang, Yu-Chung</author><author>Wang, Chen-Hao</author></authors></contributors><titles><title>Electrocatalytic activity of Nb-doped hexagonal WO3 nanowire-modified graphite felt as a positive electrode for vanadium redox flow batteries</title><secondary-title>Journal of Materials Chemistry A</secondary-title></titles><periodical><full-title>Journal of Materials Chemistry A</full-title></periodical><pages>11472-11480</pages><volume>4</volume><number>29</number><dates><year>2016</year></dates><publisher>The Royal Society of Chemistry</publisher><isbn>2050-7488</isbn><work-type>10.1039/C6TA03936G</work-type><urls><related-urls><url>http://dx.doi.org/10.1039/C6TA03936G</url></related-urls></urls><electronic-resource-num>10.1039/C6TA03936G</electronic-resource-num></record></Cite></EndNote>[3]. Sources of capacity fade is multitude, namely (i) H_2 evolution at negative electrode, (ii) V_2O_5 precipitation due to thermal instability of positive electrolyte, (iii) and the self-discharge due to the reaction between positive and negative electrolyte via membrane crossover. But, most of the organic additives are for the positive electrolyte and they are not stable in highly oxidising condition of VO_2^+. For the first time, we have explored an organic additive, D-fructose, to the negative electrolyte and arrested the capacity fade significantly ADDIN EN.CITE <EndNote><Cite><Author>Pasala</Author><Year>2017</Year><RecNum>2</RecNum><DisplayText><style Redox Balancing of Vanadium Redox Flow Battery Using D-Fructose as Negative Electrolyte Additive</title><secondary-title>ChemistrySelect</secondary-title></titles><periodical><full-title>ChemistrySelect</full-title></periodical><pages>720-727</pages><volume>2</volume><number>2</number><keywords><keyword>D-fructose</keyword><keyword>fructose crossover</keyword><keyword>hydrogen evolution</keyword><keyword>negative electrolyte</keyword><keyword>vanadium redox flow battery.</keyword></keywords><dates><year>2017</year></dates><isbn>2365-6549</isbn><urls><related-urls><url>http://dx.doi.org/10.1002/slct.201601417</url></related-urls></urls><electronic-resource-num>10.1002/slct.201601417</electronic-resource-num></record></Cite></EndNote>[4] mainly targeting H_2 evolution and self-discharge.

Recently we are studying N and P co-doped carbon derived from microcrystalline cellulose as a catalyst for the positive electrode and enhanced the EE and VE by 7.7 and 6.8 %, respectively. We have demonstrated 50 continuous cycles with minimal (< 5 %) capacity fade. In this presentation, optimization of N and P co-doping, effect of D-fructose in negative electrode kinetics and capacity retention, and flow battery performance optimization would be discussed in detail. 1.5 M VO^{2+} in 2.5 M H_2SO_4 and 1.5 M V^{3+} in 2.5 M H_2SO_4 are used as catholyte and anolyte respectively during charging in a 50 cm^2 VRFB cell. The average state-of-charge (SOC) of the VRFB was around 70 % with a cut-off voltages of 1.6 V and 0.8 V during charging and discharging, respectively. Fig. 1 compares the capacity retention for the VRFB constructed using carbon felt positive electrode, and N and P co-doped carbon based positive electrode (with and without Fructose additive used in the negative electrolyte). As can be seen, there is a tremendous improvement in the capacity retention.

Capacity retention plot of VRFB constructed using positive electrode made of (a) carbon felt, (b) N and P co-doped carbon felt, (c) N and P co-doped carbon felt with D-fructose in the negative electrolyte.

References

ADDIN EN.REFLIST 1. Jia, C.; Liu, Q.; Sun, C.-J.; Yang, F.; Ren, Y.; Heald, S. M.; Liu, Y.; Li, Z.-F.; Lu, W.; Xie, J., In Situ X-ray Near-Edge Absorption Spectroscopy Investigation of the State of Charge of All-Vanadium Redox Flow Batteries. *ACS Applied Materials & Interfaces* **2014,** *6* (20), 17920-17925.

2. Corcuera, S.; Skyllas-Kazacos, M., State-of-charge monitoring and electrolyte rebalancing methods for the vanadium redox flow battery. *European Chemical Bulletin* **2012,** *1* (12), 511-519.

3. (a) Wu, L.; Shen, Y.; Yu, L.; Xi, J.; Qiu, X., Boosting vanadium flow battery performance by nitrogen-doped carbon nanospheres electrocatalyst. *Nano Energy* **2016,** *28*, 19-28; (b) Kabtamu, D. M.; Chen, J.-Y.; Chang, Y.-C.; Wang, C.-H., Electrocatalytic activity of Nb-doped hexagonal WO3 nanowire-modified graphite felt as a positive electrode for vanadium redox flow batteries. *Journal of Materials Chemistry A* **2016,** *4* (29), 11472-11480.

4. Pasala, V.; Ramanujam, K., On In–situ Redox Balancing of Vanadium Redox Flow Battery Using D-Fructose as Negative Electrolyte Additive. *ChemistrySelect* **2017,** *2* (2), 720-727.

506402 Measurements and Calculations of Asphaltene DepositionÂ

Abstract:

Asphaltene precipitation and deposition have been recognized as curtailing and vital problems in oil and gas fields. Asphaltene precipitation and deposition affect production, transmission and processing facilities in different and tremendous ways. These problems lead to system interruptions and have become a crucial part of flow assurance studies. One way to mitigate these problems is to keep the production system free of wax and asphaltenes.

Different approaches have been implemented in the oil and gas industries to predict asphaltene precipitation or deposition conditions. Predicting the asphaltene onset pressure (AOP) requires the presence of pressurized down-hole samples with very low or no oilâ€"based mud contamination, which is not an easy task.

Several models have been introduced in the oil industry; some of them are considered costly and time consuming since they require experimental work to come up with an accurate prediction.

The asphaltene precipitation conditions in Kuwait oil fields from both South East Kuwait (SEK) and West Kuwait (WK) were studied extensively to develop a fast and low-cost model with acceptable accuracy.

This study has three phases. The first phase is a literature review regarding the available approaches to predict the AOP. Later, a new model is introduced to overcome the current industrial obstacles. The new model, which relies on basic reservoir data such as the temperature, pressure, bubble point and a SARA analysis, was tested using lab measurements.

In the final stage, PIPESIM software was used to predict the AOP for eighteen Kuwaiti oil wells. The accuracy of the proposed model as well as the PIPESIM predictions indicated that both approaches resulted in some error compared with the measured values of the AOP. The model developed in this study has an average absolute error of 12% compared to 7.50% by PIPESIM. However, the proposed model is easy to be used and does not need commercial license.

512068 Research on Solution of Suspending Seabed Pipeline Based on Bionic Protective Technology

Abstract:

The paper intends to propose an immediate and also permanent solution to suspending seabed pipeline, which is a fatal factor of the oil spill accident, caused by scour of currents, instability of the seabed, vibration of the riser and others developed by the present authors. Based on the marine dynamics and marine bionics, the mechanism and effect of scour is introduced and a new protective technique is explored. As core of technique, composite material fronts mats are secured to the seabed using anchors and the buoyant fronts reduce the current velocity and sediment transportation to protect the pipeline rely on viscous drag, causing sediment to fall into the fronds to back fill scour pits and to form a permanent reinforced sandbank. Compared with other traditional methods, which seldom prohibit second scour, the efficiency and feasibility of this technique could be guaranteed. In order to demonstrate the advancement, a set of model experiments are provided to show the status of deposition with time. Three cases of application in risers of an oilfield company indicate that the protective system with the sediment deposition has meaningful influence on elimination and rectification scour. With the characters of "one time", "one cost" installation, low installed cost, environmental friendly material and cheap maintaining, this protective technique has benefit to provide a permanent, self-maintaining solution to scour, improve of the integrity of the suspending seabed pipeline ,finally to control oil spill accident.

512539 A Heat-Recirculating Combustor Using SiC Porous Foam for Thermophotovoltaic Power Devices

Abstract:

Efficient and reliable portable power devices have been investigated due to the increasing demand of electronic devices. Because of the advantage of the high energy density, power systems using the combustion of hydrocarbon fuels are considered as one alternative to replace secondary batteries. A thermophotovoltaic (TPV) power system, which is one of the combustion-driven power systems, converts thermal energy to electricity through photons. The TPV power systems consist of simple structure and have high efficiency without friction loss by moving parts. To enhance the efficiency of the TPV systems, several design factors are important: stable burning in the combustor as a heat source and high enough temperature of the emitter as a radiation source. In this study, to improve the performance of the combustor that was developed for TPV power devices in a previous study, silicon carbide (SiC) porous foam is inserted into a combustion chamber. Porous burners have been considered as one possible technology for achieving the high-efficiency by increasing residence time for complete combustion and enhancing internal heat recirculation. To evaluate the combustor performance, combustion stability limits and temperature distribution along the emitter wall surface are measured. Compared with the case without SiC porous foam in the combustion chamber, the flame stability limits are extended with SiC porous foam, implying a widened operating range of the combustor. The temperature of the emitter wall surface is locally enhanced where SiC porous foam is installed, and the average temperature of the emitter wall also increases when SiC porous foam is used.

513034 Combustion Properties of Outwardly-Propagating Spherical CH4-NH3/Air Flames

Abstract:

Fundamental combustion characteristics of methane (CH_4)-ammonia (NH_3) blends are studied to evaluate the potential as a low-carbon fuel since their utilization can be considered as an interim technology for responding to global warming. The propagation of spark-ignited spherical laminar premixed CH_4-NH_3/air flames at normal temperature and pressure is experimentally investigated and also computationally predicted using a detailed kinetic mechanism. The fundamental unstretched laminar burning velocities and Markstein numbers are determined and the associated flame structure is investigated. Results show that the laminar burning velocities are reduced with NH_3 substitution, while their tendency with varying fuel-equivalence ratio is similar to the pure CH_4/air flames. Predicted flame structure exhibits that nitrogen oxide (NO_x) formation is enhanced with NH_3 substitution, but at fuel-rich conditions the increased amount of NO_x emissions is much lower than that under fuel-lean conditions. Markstein numbers of CH_4-NH_3/air flames are similar to the pure CH_4/air flames, implying that the flamefront instabilities are not significantly affected by NH_3 substitution. Measured and predicted tendencies of the laminar burning velocities and Markstein numbers for various conditions show encouraging agreement, but quantitative discrepancies among the measurements and predictions at some conditions merit additional consideration.

511675 Implementation of Distributed Co-Simulation for Robust Design of Urban Energy Systems

Abstract:

Simulation can support an optimal design and operation strategy of multi-energy systems. One approach being co-simulation allows the use of dedicated software and tools for each subsystem.

Except for tools offering Functional Mock-up Interface (FMI) functionalities, there is no common standardized interface that can be used for the co-simulation of such multi-energy systems. Rather than trying to create new standards, the goal here is to create a robust and adaptable co-simulation platform, acting as a easily linkable backbone.

This presentation is the opportunity to show how we implemented a distributed multi-platform co-simulation using containerization concept and an orchestrator to link tools such as TRNSYS, EnergyPlus, Modelica, Matlab and Python based tools, with or without FMI.

Currently a lack of robust software implementation skills can lead to misused, unused or unmaintainable academic tools. The presented platform will enable non computer specialists to exploit the possibilities of co-simulation techniques while respecting standards for tool development that will make their contributions reusable by others and easily maintainable.

We will present the application of the co-simulation platform on a use case based on the city of Vevey (Switzerland) combining a low temperature district heating and cooling network with decentralized heat pumps, an electricity network with power to gas technologies and a gas powered combined heat and power production plant supplying a high temperature district heating network and compare the use of the distributed co-simulation platform with that of a classical monolithic simulation using Modelica.

This method is applied for the implementation of distributed multi-platform co-simulation for hybrid urban energy systems in the IntegrCiTy project funded by the ERA-NET Cofund Smart Cities and Communities.

512947 The Feasibility of Replacing Electric Drives with Wind Energy in the Swro System

Abstract:

The feasibility of substituting electricity in the Sea Water Reverse Osmosis (SWRO) process with wind energy is considered in this article.

The SWRO process is divided into very high and medium pressure areas. The medium pressure areas are listed below;

1. Sea water cleaning and conditioning section
2. High purity water conditioning
3. Product water export

The hydraulic system for the medium pressure areas consists of pipe networks that supply Pelton or turbine drives (P-drives) for pumps. The energy input to the hydraulic network is from wind turbine driven pumps.

The very high pressure area of the SWRO unit is a once through system. A wind turbine-pump pair capable of producing the required head is considered.

512962 Bio-Drying: A Sustainable Approach for Enhancing the Pre-Treatment Process and Thermal Properties of Green Waste

Abstract:

Abstract

The potential for thermal recovery of waste is gaining impetus among researchers and industries across the globe especially in many developed countries. However, in processing waste for energy recovery process, the type and nature of input waste materials, particularly those with high moisture content, have a significant impact in determining the quality, environmental profile of the waste technology as well as the thermal properties of the final product. Bio-drying, as a waste to energy conversion technology, tends to reduce moisture content of waste while maintaining the energy content of the processed waste. The current study investigates the effect of input materials (plastics, sorted waste papers, pruning waste and food waste) on the energy content of waste material with subsequent implication on biogenic content of the material by bio-drying process. Preliminary results established a positive correlation between biogenic and energy content of the input materials with some variations observed between food and paper waste. High biogenic content was reported for plastic waste whereas paper waste had the highest non-biogenic content of 9%. Further analysis after the bio-drying process show that, depending on the proportion of the different waste materials, the biogenic and energy content of the final material differed indicating the significant of the amount of some materials in the waste mix with respect to others. Among the waste components plastic was identified as the major contributor to high biogenic and energy content, whereas paper was considered as the lowest contributor. Additionally, high proportion of food waste in the waste mix indicated a slight difference in biogenic and energy content; on the other hand, the same proportion of paper in the waste mix showed similar biogenic content with some variation of energy content observed. This research revealed that the combination of green waste of high moisture content with other waste materials by bio-drying process would offer alternative waste management approach and help mitigate environmental pollution from the disposal of biodegradable waste.

Keywords: Bio-drying; Biogenic content; Energy content;

Solar Thermochemical H_2O/CO_2 Splitting Using Ni-based Ferrite Materials

(Principal Author)
Rahul Bhosale
Department of Chemical Engineering
College of Engineering
Qatar University, Doha, Qatar
rahul.bhosale@qu.edu.qa

Prepared for Presentation at
American Institute of Chemical Engineers
2018 Spring Meeting
14th Global Congress on Process Safety
Orlando, Florida
April 22-26, 2018

Solar Thermochemical H_2O/CO_2 Splitting Using Ni-based Ferrite Materials

Rahul Bhosale
Department of Chemical Engineering
College of Engineering
Qatar University, Doha, Qatar
rahul.bhosale@qu.edu.qa

Keywords: Ni-ferrite, H_2O/CO_2 Splitting, Solar Energy, Thermodynamics

Abstract

In this paper, a thermodynamic analysis of Ni-ferrite based H_2O splitting cycle is performed using HSC Chemistry 7.1 software and its thermodynamic database. The total amount of solar energy required to run the cycle ($\dot{Q}_{solar-cycle}$), total amount of heat energy re-radiated from the cycle ($\dot{Q}_{re-rad-cycle}$), amount of heat energy released by water splitting reactor ($\dot{Q}_{splitting-reactor}$), cycle ($\eta_{cycle}$) and solar-to-fuel energy conversion efficiency ($\eta_{solar-to-fuel}$), and other crucial thermodynamic parameter are determined at T_H = 2000 K and T_L = 1200 K (P_{O_2}= 10^{-4} atm).

1. Introduction

Solar-driven thermochemical splitting of H_2O/CO_2 using metal oxides is one of the alternatives for the production of H_2 and syngas[1–35]. The H_2 can be used directly as a transportation fuel, and syngas can be processed further for the manufacturing of liquid transportation fuel via Fischer Tropsch process. Ferrites are one of the class of metal oxides which has been used significantly for the production of H_2 via water splitting reaction. Among the various ferrites investigated until now, Ni-ferrite is one of the best materials reported by the investigators. It can produce higher amounts of H_2 as compared to other ferrites with repeated cyclability. The redox reactions involved in case of Ni-ferrite based H_2O splitting cycle are as follows:

$$NiFe_2O_4 \rightarrow NiFe_2O_{4-\delta} + {}^{\delta}/_{2}\, O_2 \quad (1)$$
$$NiFe_2O_{4-\delta} + \delta H_2O \rightarrow NiFe_2O_4 + \delta H_2 \quad (2)$$

In this paper, the thermodynamic analysis of solar H_2 production via Ni-ferrite based H_2O splitting cycle is performed using HSC Chemistry 7.1. The principles of the second law of thermodynamics are applied and the solar-to-fuel energy conversion efficiency for the Ni-ferrite based H_2O splitting cycle is determined. The thermal reduction is carried out at 2000 K (by applying partial pressure of O_2 in the inert gas equal to 10^{-4} atm). Also, the water-splitting reaction is performed at 1200 K. All the thermodynamic calculations are performed by using the process flow diagram presented in Figure 1 and by considering molar flowrate of Ni-ferrite equal to 1 mol/s entering the solar thermal reactor.

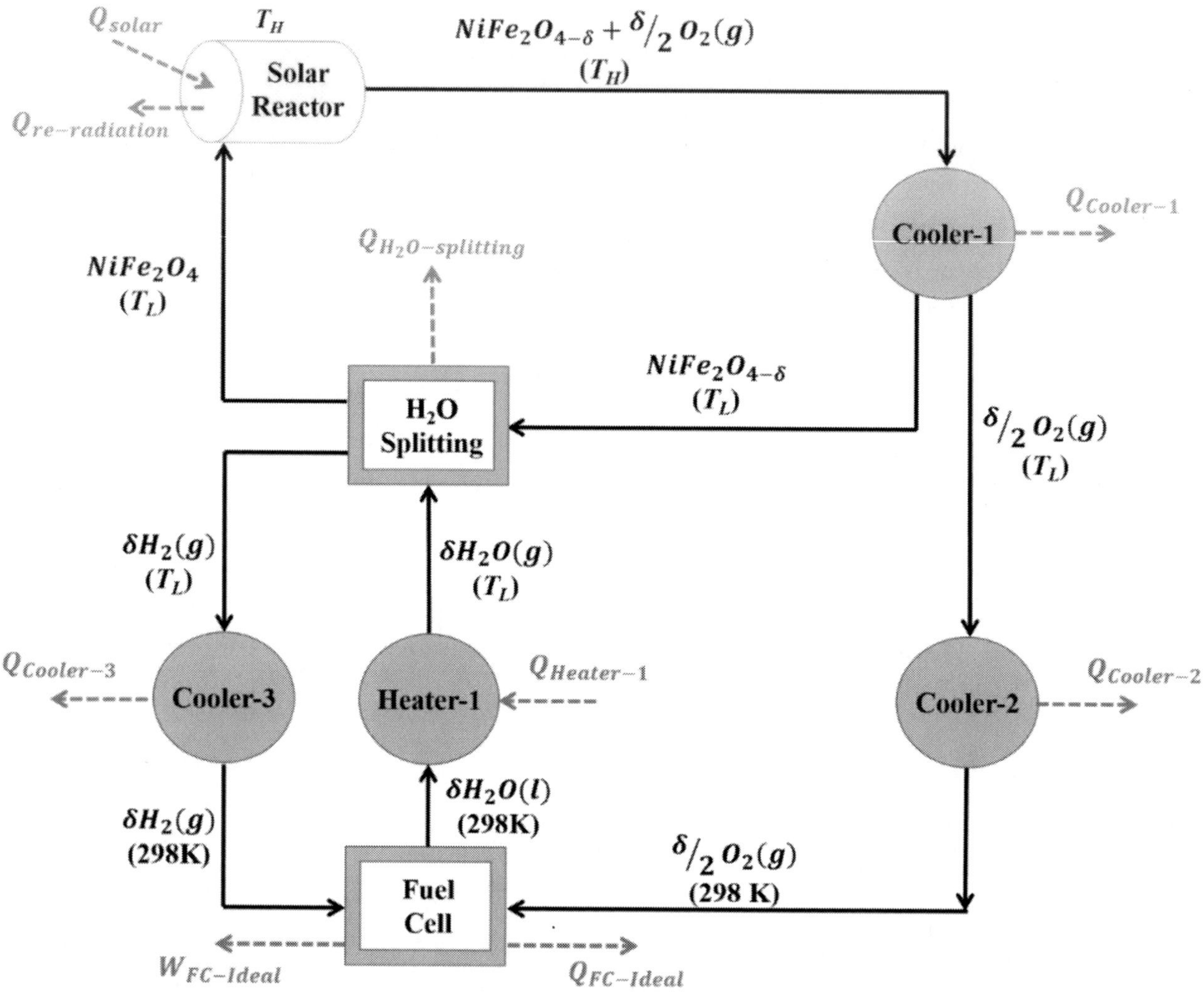

Figure 1. Ni-ferrite based water splitting cycle (process flow diagram used for the thermodynamic analysis).

2. Solar Reactor Efficiency Analysis

According to the thermodynamic equilibrium analysis, 0.4925 mol of O_2 is released from Ni-ferrite when the thermal reduction is carried out at 2000 K (P_{O_2}= 10^{-4} atm). It is important to note that the thermal reduction of $NiFe_2O_4$ results in the production of a solid solution containing Ni^{+2}, Fe^{+2}, and Fe^{+3} oxides. These oxides are combined to determine the δ which yields into the final form of Ni-ferrite as $NiFe_2O_{4-\delta}$.

The net amount of heat energy desirable to run the Ni-ferrite based H_2O splitting cycle is calculated as per the following equation.

$$\dot{Q}_{cycle-net} = \dot{Q}_{NiFe_2O_4-red} + \dot{Q}_{H_2O-heating} \quad (3)$$

Where,

$$\dot{Q}_{NiFe_2O_4-red} = \dot{n}\Delta H|_{NiFe_2O_4@T_L \rightarrow NiFe_2O_{4-\delta} + \delta/_2 O_2(g)@T_H} \quad (4)$$

$$\dot{Q}_{H_2O-heating} = \dot{n}\Delta H|_{\delta H_2O(l)@298K \rightarrow \delta H_2O(g)@T_L} \quad (5)$$

At T_H = 2000 K and T_L = 1200 K (P_{O_2}= 10^{-4} atm), the $\dot{Q}_{H_2O-heating}$ = 71.1 kW, $\dot{Q}_{NiFe_2O_4-red}$ = 671.8 kW, and $\dot{Q}_{cycle-net}$ = 742.9 kW.

The total amount of solar energy necessary to drive the Ni-ferrite based H_2O splitting cycle is determined as:

$$\dot{Q}_{solar-cycle} = \dot{Q}_{solar-reactor} + \dot{Q}_{solar-heater} \quad (6)$$

Where,

$$\dot{Q}_{solar-reactor} = \frac{\dot{Q}_{NiFe_2O_4-red}}{\eta_{abs-solar-reactor}} \quad (7)$$

$$\dot{Q}_{solar-heater} = \frac{\dot{Q}_{H_2O-heating}}{\eta_{abs-solar-heater}} \quad (8)$$

$$\eta_{abs-solar-reactor} = 1 - \left(\frac{\sigma T_H^4}{IC}\right) \quad (9)$$

$$\eta_{abs-solar-heater} = 1 - \left(\frac{\sigma T_L^4}{IC}\right) \quad (10)$$

At T_H = 2000 K and T_L = 1200 K (P_{O_2}= 10^{-4} atm), the $\eta_{abs-solar-heater}$ = 96.1%, $\eta_{abs-solar-reactor}$ = 69.8%, $\dot{Q}_{solar-heater}$ = 74.1 kW, $\dot{Q}_{solar-reactor}$ = 963.0 kW, and $\dot{Q}_{solar-cycle}$ = 1037.1 kW, respectively.

The re-radiation losses from the Ni-ferrite based H_2O splitting cycle are estimated as:

$$\dot{Q}_{re-rad-cycle} = \dot{Q}_{re-rad-solar-reactor} + \dot{Q}_{re-rad-solar-heater} \quad (11)$$

Where,

$$\dot{Q}_{re-rad-solar-reactor} = \dot{Q}_{solar-reactor} - \dot{Q}_{NiFe_2O_4-red} \quad (12)$$

$$\dot{Q}_{re-rad-solar-heater} = \dot{Q}_{solar-heater} - \dot{Q}_{H_2O-heating} \quad (13)$$

At T_H = 2000 K and T_L = 1200 K (P_{O_2}= 10^{-4} atm), the $\dot{Q}_{re-rad-solar-heater}$ = 2.91 kW, $\dot{Q}_{re-rad-solar-reactor}$ = 291.3 kW, and $\dot{Q}_{re-rad-cycle}$ = 294.2 kW.

The heat dissipated by the three coolers is gauged as per the following equations:

$$\dot{Q}_{cooler-1} = -\dot{n}\Delta H|_{NiFe_2O_{4-\delta} + \delta/_2 O_2(g)@T_H \rightarrow NiFe_2O_{4-\delta} + \delta/_2 O_2(g)@T_L} \quad (14)$$

$$\dot{Q}_{cooler-2} = -\dot{n}\Delta H|_{\delta/_2 O_2(g)@T_L \rightarrow \delta/_2 O_2(g)@298K} \quad (15)$$

$$\dot{Q}_{cooler-3} = -\dot{n}\Delta H|_{H_2(g)@T_L \rightarrow H_2(g)@298K} \quad (16)$$

At T_H = 2000 K and T_L = 1200 K (P_{O_2}= 10^{-4} atm), the $\dot{Q}_{cooler-1}$ = 213.6 kW, $\dot{Q}_{cooler-2}$ = 14.7 kW, and $\dot{Q}_{cooler-3}$ = 26.4 kW.

The heat energy released during water splitting reactor is also calculated as per Eq. (17) and observed to be equal to 431.9 kW.

$$\dot{Q}_{splitting-reactor} = -\dot{n}\Delta H|_{NiFe_2O_{4-\delta} + \delta H_2O(g)@T_L \rightarrow NiFe_2O_4 + \delta H_2(g)@T_L} \quad (17)$$

The highest amount of work that can be obtained and the heat energy that will be liberated from the ideal fuel cell is determined as:

$$\dot{W}_{FC-Ideal} = -\dot{n}\Delta G|_{\delta H_2(g) + \delta/_2 O_2(g)@298K \rightarrow \delta H_2O(l)@298K} \quad (18)$$

$$\dot{Q}_{FC-Ideal} = -(298) \times \dot{n}\Delta S|_{\delta H_2(g) + \delta/_2 O_2(g)@298K \rightarrow \delta H_2O(l)@298K} \quad (19)$$

At T_H = 2000 K and T_L = 1200 K (P_{O_2}= 10^{-4} atm), the $\dot{W}_{FC-Ideal}$ = 233.6 kW, and $\dot{Q}_{FC-Ideal}$ = 47.9 kW.

The cycle and solar-to-fuel energy conversion efficiency for the Ni-ferrite based H_2O splitting cycle are computed as:

$$\eta_{cycle} = \frac{\dot{W}_{FC-Ideal}}{\dot{Q}_{solar-cycle}} \quad (20)$$

$$\eta_{solar-to-fuel} = \frac{\delta \times HHV_{H_2}}{\dot{Q}_{solar-cycle}} \quad (21)$$

At T_H = 2000 K and T_L = 1200 K (P_{O_2}= 10^{-4} atm), the η_{cycle} = 22.5%, and $\eta_{solar-to-fuel}$ = 27.1%.

4. Conclusion

The Ni-ferrite based water splitting cycles is thermodynamically studied to determine the various thermodynamic parameters associated with the solar reactor design. This thermodynamic calculation is performed at T_H = 2000 K and T_L = 1200 K (P_{O_2}= 10^{-4} atm). The various thermodynamic parameters are as follows: $\dot{Q}_{H_2O-heating}$ = 71.1 kW, $\dot{Q}_{NiFe_2O_4-red}$ = 671.8 kW, and $\dot{Q}_{cycle-net}$ = 742.9 kW, $\eta_{abs-solar-heater}$ = 96.1%, $\eta_{abs-solar-reactor}$ = 69.8%, $\dot{Q}_{solar-heater}$ = 74.1 kW, $\dot{Q}_{solar-reactor}$ = 963.0 kW, $\dot{Q}_{solar-cycle}$ = 1037.1 kW, $\dot{Q}_{re-rad-solar-heater}$ = 2.91 kW, $\dot{Q}_{re-rad-solar-reactor}$ = 291.3 kW, $\dot{Q}_{re-rad-cycle}$ = 294.2 kW, $\dot{Q}_{cooler-1}$ = 213.6 kW, $\dot{Q}_{cooler-2}$ = 14.7 kW, and $\dot{Q}_{cooler-3}$ = 26.4 kW, $\dot{W}_{FC-Ideal}$ = 233.6 kW, $\dot{Q}_{FC-Ideal}$ = 47.9 kW, $\dot{Q}_{splitting-reactor}$ = 431.9 kW, η_{cycle} = 22.5%, and $\eta_{solar-to-fuel}$ = 27.1%.

5. Acknowledgments

This publication was made possible by the NPRP grant (NPRP8-370-2-154) from the Qatar National Research Fund (a member of Qatar Foundation). The statements made herein are solely the responsibility of author(s).

References

(1) Agrafiotis, C. C.; Pagkoura, C.; Lorentzou, S.; Kostoglou, M.; Konstandopoulos, A. G. *Catal. Today* **2007**, *127* (1–4), 265–277.
(2) Kodama, T.; Nakamuro, Y.; Mizuno, T. *J. Sol. Energy Eng.* **2006**, *128* (1), 3.
(3) Gokon, N.; Murayama, H.; Umeda, J.; Hatamachi, T.; Kodama, T. *Int. J. Hydrogen Energy* **2009**, *34* (3), 1208–1217.
(4) Miller, J. E.; Allendorf, M. D.; Diver, R. B.; Evans, L. R.; Siegel, N. P.; Stuecker, J. N. *J. Mater. Sci.* **2008**, *43* (14), 4714–4728.
(5) Bhosale, R. R.; Kumar, A.; Van Den Broeke, L. J. P.; Gharbia, S.; Dardor, D.; Jilani, M.; Folady, J.; Al-Fakih, M. S.; Tarsad, M. A. *Int. J. Hydrogen Energy* **2015**, *40* (4), 1639–1650.
(6) Steinfeld, A. *Int. J. Hydrogen Energy* **2002**, *27* (6), 611–619.
(7) Abanades, S. In *Industrial and Engineering Chemistry Research*; 2012; Vol. 51, pp 741–750.
(8) Dardor, D.; Bhosale, R.; Gharbia, S.; Kumar, A.; AlMomani, F. *J. Emerg. Trends Eng. Appl. Sci.* **2015**, *6* (September 2016), 129–135.
(9) Koepf, E.; Villasmil, W.; Meier, A. *Appl. Energy* **2016**, *165*, 1004–1023.
(10) Abanades, S.; Charvin, P.; Lemont, F.; Flamant, G. *Int. J. Hydrogen Energy* **2008**, *33* (21), 6021–6030.
(11) Charvin, P.; Abanades, S.; Lemont, F.; Flamant, G. *AIChE J.* **2008**, *54* (10), 2759–2767.
(12) Bhosale, R. R.; Kumar, A.; Sutar, P. *Energy Convers. Manag.* **2017**, *135*, 226–235.
(13) Lorentzou, S.; Agrafiotis, C. C.; Konstandopoulos, A. G. *Granul. Matter* **2008**, *10* (2), 113–122.
(14) Scheffe, J. R.; Li, J.; Weimer, A. W. *Int. J. Hydrogen Energy* **2010**, *35* (8), 3333–3340.

(15) Muhich, C. L.; Aston, V. J.; Trottier, R. M.; Weimer, A. W.; Musgrave, C. B. *Chem. Mater.* **2016**, *28* (1), 214–226.

(16) Muhich, C. L.; Evanko, B. W.; Weston, K. C.; Lichty, P.; Liang, X.; Martinek, J.; Musgrave, C. B.; Weimer, A. W. *Science (80-.).* **2013**, *341* (6145), 540–542.

(17) Bhosale, R. R.; Shende, R. V.; Puszynski, J. A. *MRS Proc.* **2012**, *1387*, mrsf11-1387-e09-07.

(18) Bhosale, R. R.; Kumar, A.; AlMomani, F.; Ghosh, U.; Sutar, P.; Takalkar, G.; Ashok, A.; Alxneit, I. *Ceram. Int.* **2017**, *43* (6), 5150–5155.

(19) Bhosale, R. R.; Shende, R. V.; Puszynski, J. A. *Int. J. Hydrogen Energy* **2012**, *37* (3), 2924–2934.

(20) Bhosale, R. R.; Alxneit, I.; van den Broeke, L. L. P.; Kumar, A.; Jilani, M.; Gharbia, S. S.; Folady, J.; Dardor, D. Z. *MRS Proc.* **2014**, *1675*, mrss14-1675-rr06-10.

(21) Bhosale, R. R.; Kumar, A.; Almomani, F.; Alxneit, I. *Ceram. Int.* **2016**, *42* (2), 2431–2438.

(22) Abanades, S.; Flamant, G. *Sol. Energy* **2006**, *80* (12), 1611–1623.

(23) Ackermann, S.; Scheffe, J. R.; Steinfeld, A. *J. Phys. Chem. C* **2014**, *118* (10), 5216–5225.

(24) Bhosale, R. R.; Kumar, A.; Almomani, F.; Ghosh, U.; Al-Muhtaseb, S.; Gupta, R.; Alxneit, I. *Ceram. Int.* **2016**, *42* (8), 9354–9362.

(25) Lin, F.; Rothensteiner, M.; Alxneit, I.; van Bokhoven, J. A.; Wokaun, A. *Energy Environ. Sci.* **2016**, 2400–2409.

(26) Bhosale, R. R.; Kumar, A.; AlMomani, F.; Alxneit, I. *Ceramics International*. 2015.

(27) Ackermann, S.; Sauvin, L.; Castiglioni, R.; Rupp, J. L. M.; Scheffe, J. R.; Steinfeld, A. *J. Phys. Chem. C* **2015**, *119* (29), 16452–16461.

(28) Chueh, W. C.; Abbott, M.; Scipio, D.; Haile, S. M. *Science (80-.).* **2010**, *63* (December), 2010.

(29) Scheffe, J. R.; Weibel, D.; Steinfeld, A. In *Energy and Fuels*; 2013; Vol. 27, pp 4250–4257.

(30) Nair, M. M.; Abanades, S. *ChemistrySelect* **2016**, *1* (15), 4449–4457.

(31) Carrillo, R. J.; Scheffe, J. R. *Solar Energy*. 2017.

(32) Dey, S.; Rao, C. N. R. *ACS Energy Lett.* **2016**, *1* (1), 237–243.

(33) Dey, S.; Naidu, B. S.; Govindaraj, A.; Rao, C. N. R. *Phys. Chem. Chem. Phys.* **2014**, *17* (1), 122–125.

(34) Takacs, M.; Hoes, M.; Caduff, M.; Cooper, T.; Scheffe, J. R.; Steinfeld, A. *Acta Mater.* **2016**, *103*, 700–710.

(35) Rahul R. Bhosale, Anand Kumar, Anchu Ashok, Parag Sutar, Gorakshnath Takalkar, Majeda Khraisheh, Fares AlMomani, U. G. *MRS Adv.* **2017**, No. May, 1–8.

513951 Recovering Hydrogen Biofuel from Wet-Biowaste

Abstract:

Problems with hydrogen and biofuel:

Most fossil-based hydrogen is produced by steam methane reforming (SMR) and renewable hydrogen from the electrolysis of water. Without considering plant efficiency, capital or operating costs, hydrogen from SMR has a minimum value of 1.45 times the price of methane, a valuable and resource-limited commodity. In addition, for every ton of hydrogen produced, 10 ton of greenhouse-gas carbon dioxide is co-produced and released from the feedstock and in processing. The electrolysis of water requires a clean, pure feedstock and requires considerably more energy than what hydrogen will produce when reacted with oxygen (air).

In contrast to inorganic hydrogen production as described above, organic biofuels are produced by anaerobic digestion and fermentation, which are proven and well-established technologies. Anaerobic digestion uses decaying organic material to produce a syngas where there is waste that needs disposal, or significant subsidies for using energy crops. Fermentation has been producing alcohol beverages for thousands of years from cultivated crops, now it's used to produce subsidized ethanol from corn and sugar, impacting feed and food costs. To quiet the "food vs. fuel" debate, effort is now directed at cultivating algae and grass feedstock. Cultivation requires fossil-fuel for planting, fertilizing, harvesting, transporting and distillation of product, which without subsidies would be highly unprofitable.

Anaerobic digestion and fermentation both use biological microorganisms ('bugs'), which are temperature dependent, require large volumes with slow rates & low-yields, and are prone to contamination (sulfur) or conditions (temperature) that kills the 'bugs' employed to do the work. In addition, conventional biofuel producers must pay for crop feedstocks whereas biowaste producers pay for biowaste disposal; yet, biowaste has higher energy content than crop feedstocks.

An alternative, recovering hydrogen biofuel from wet-biowaste:

In the U.S., waste-water treatment plants (WWTP), municipal solid waste, confined area feeding operations, and agriculture produce over 1 billion tons of organic biowaste annually that can cost from $40-$200 per ton to treat before use or disposal. In addition, the amount and cost is increasing.

Fuel and electricity represent a substantial cost to WWTPs; one is required for nearly all stages in the treatment process, from the collection of raw sewage to the discharge of treated effluent. An estimated 3-4% of total U.S. electricity consumption is used for the transfer and treatment of wastewater, corresponding to more than 45 million tons of greenhouse gas (GHG) emissions annually.

WWTP in the U.S. produced 7.9 million tons of sewage biowaste in 2014. The latent energy in this biowaste could meet 12% of the nation's electricity demand. A 10 million gallon/day WWTP producing 7 tons/day of biowaste will require 11 to 17 MWhe of electricity daily. In addition, due to water shortages, higher energy & capital costs and a changing climate, water-energy issues are of growing importance in biowaste management at WWTP.

In contrast to being a public energy, economic and environmental burden, Chemergy's patented HyBrTec technology allows WWTPs to be energy independent and profitable producers of thermal energy and renewable fuel. All organic biowaste contains an unused remnant of stored solar energy that HyBrTec recovers as renewable hydrogen, non-anthropogenic carbon dioxide, thermal energy and inorganic residuals suitable as a micro-nutrient fertilizer.

Processing wet-biowaste, HyBrTec exploits two well-established steps that are scalable from pounds to tons per minute with commercially available components. First, the wet-biowaste ($C_6H_{10}O_5$ & H_2O) is 'burned' or oxidized with bromine (Br_2) to produce hydrogen bromide (HBr), carbon dioxide (CO_2) and heat ($C_6H_{10}O_5 + 7H_2O + 12Br2 \rightarrow 24HBr + 6CO_2$). The HBr reacts with unreacted water forming concentrated hydrobromic acid (HBr_{aq}). Bromination is followed by electrolysis of the hydrobromic acid into hydrogen (H_2) and reagent Br_2 and dilute acid that is recycled back into the process.

Hydrogen bromide electrochemistry is reversible ($2HBr \leftrightarrow H_2 + Br_2$) allowing the charging (storing) and discharging (production) of electrical energy. This enables a HyBrTec reversible electrolysis system to co-provide efficient electrical energy storage. The hydrogen-bromine energy storage system operates as a flow battery where the reactants are stored outside the electrochemical PEM stack, which simplifies capacity growth. Energy storage increases the value of off-peak and intermittent solar & wind energy and promotes development of micro- and smart-grids that can exploit local sources of energy, improve efficiency, reduce needed capacity and mitigate disruptions.

Feedstocks include: sewage, manure, agricultural residuals, and municipal solid waste including plastics. Also, wastewater is a reagent in the process producing additional hydrogen. Unique to HyBrTec is that 175-200 C heat (26.3 kWh_{th}/kgH_2) is released, which can be used to reduce the feedstock water content to a desired 50%. Also, electroosmotic water transfer from the bromine anode to the hydrogen cathode increases the acid concentration at the anode, which lowers cell voltage and results in 4-6 gallons of potable water per kgH_2.

If the residual energy content of the biowaste is omitted, HyBrTec offers a *theoretical* biowaste-to-hydrogen-to-energy efficiency greater than 100%. This is simply because the theoretical voltage to dissociate hydrogen carrier hydrogen bromide into hydrogen and bromine (0.58 Volt) is less than the voltage produced in the hydrogen reaction with oxygen (1.23 Volt) forming water, providing a *theoretical* biowaste-to-energy efficiency of 212% (1.23V/0.58V). However, if the electrolyzer operates at an actual 0.75-0.8 Volts (16-21.23 kWh/kgH_2 depending on electrolyte temperature) the efficiency is 154%, and if the hydrogen is used in a 65% efficient fuel cell with air the electric-to-electric efficiency is reduced further to 100%. The efficiencies above disregard the latent energy content of the negative-valued feedstock; however, if included the biowaste-to-electricity efficiency is a healthy 67%.

Hydrogen can enrich natural gas to increase its Btu content and reduce NOx emissions, used directly as a clean fuel or sold as a commodity. Furthermore, byproduct carbon dioxide is organic in origin and not considered a GHG, can be vented, or used with recovered hydrogen to synthesize methanol, ethanol or methane using Fischer-Tropsch or the Audi e-gas process. Also, recovered hydrogen can be combined with nitrogen from on-site membrane air-separation to produce ammonia, which is a higher-value commodity and easily transportable. Other opportunities to monetize the carbon dioxide byproduct include reacting with ammonia to produce even higher value urea.

Beginning in 2010 the U.S. Department of Energy and the Florida Hydrogen Initiative funded laboratory bench-top R&D of HyBrTec. Based on the success of this research and at the recommendation of Lawrence Livermore National Laboratory (LLNL), the California Energy Commission (CEC) in 2013 funded Chemergy in a cost-shared program to determine suitability of biosolids from the Delta Diablo WWTP, prepare a detailed system design & cost estimate of a demonstration system and an economic analysis of a HyBrTec commercial system.

The results of the CEC program verified that 21.44 kg of hydrogen could be produced per wet-ton of sewage sludge and providing a 95% reduction of the wet-mass. An economic analysis for a small system processing 11 wet-tons of sewage sludge per day into 236 kg of hydrogen predicted a hydrogen production cost of $2/kg and a 14% Internal Rate of Return. Economic benefits from an increased scale, energy storage, state & federal renewable energy tax incentives, loan guarantees, cap & trade credits and subsidies were not included in the analysis, which if taken would increase the IRR.

Summary:

HyBrTec offers advantages over conventional hydrogen and biofuel production:

- Wet-biowaste is abundant and an environmental and economic burden.
- The technology exploits two advantages that reduce capital and energy:
- At moderate temperature & pressure, processing is fast and yields are high, which minimizes the size, footprint and cost of equipment.
- The chemical bonds to release hydrogen are weak, requiring less than half the energy (40%) that hydrogen will produce when burned with oxygen (air).
- HyBrTec is a highly scalable technology able to process lb/day or tons/minute with commercially available components anywhere biowaste is produced.
- With electricity from renewable solar & wind resources, HyBrTec is GHG neutral.
- Hydrogen and by-product carbon dioxide can produce conventional fuels including ethanol, methane, green diesel or higher-valued chemical commodities.
- HyBrTec offers a profitable triad: 1) eliminates biowaste, 2) low-cost ($2/kg) renewable hydrogen and 3) co-provides efficient energy storage.
- $2/kg hydrogen fueling a 50% efficient fuel cell vehicle is equivalent to gasoline at $1/gallon without the harmful 'well-to-wheels' environmental and health effects.
- Accrues benefits from state and federal renewable-energy tax incentives, loan guarantees, GHG cap & trade programs and energy storage subsidies.

In closing, Chemergy's HyBrTec technology promotes a highly advantageous paradigm shift departing from: 1) costly non-renewable resources to negative-valued renewable resources, 2) large, expensive centralized plants to low-cost, well-distributed systems, 3) collection, transporting & disposal, to on-site utilization.

ZINC CARBONATE MODIFICATION OF METHANOL SYNTHESIS CATALYST FOR SINGLE-STAGE DME PRODUCTION

Lujie Ye and Sunggyu Lee
Sustainable Energy and Advanced Materials (SEAM) Laboratory
Ohio University
Athens, OH

Extended Abstract

For the catalytic synthesis of methanol from syngas as well as the optional co-production of methanol and dimethyl ether, the conventional $Cu/ZnO/Al_2O_3$ catalyst has been proven to be effective for the feed syngas containing a CO_2 concentration between 4-6 vol.%. However, if the feed syngas contains a higher amount of CO_2 such as >8 vol.%, the catalytic reactivity would dramatically decrease [1]. This phenomenon may be called "a lack of the CO_2 tolerance by the catalyst". A carbonate modified catalyst of $Cu/ZnCO_3/Al_2O_3$ developed by Lee and coworkers with the post-reduction modification approach showed an improved CO_2 tolerance in the CO_2-rich syngas environment [2, 3, 4]. While this reformulation idea has provided a needed enhancement, its post-reduction treatment on the activated catalyst has hampered its practical application in industrial reactors.

In this work, a supercritical pre-reduction modification approach (SPMA) was conceived and developed to reformulate a commercially available oxide-based catalyst before its reduction treatment, i.e., before the activation of the catalyst. With this novel approach, a carbonate modified dual-catalytic system between $CuO/ZnCO_3/Al_2O_3$ and γ-Al_2O_3 was successfully devised and created using the SEAM's proprietary condensed-phase reactive modification technology. The current concept of pre-reduction modification of the catalyst enables the modified catalyst to be commercially mass-producible in a stabilized form, which was not possible with the post-reduction modification approach.

The SPMA process is typically carried out in a stainless steel autoclave-based reactor system. Chemically pure CO_2 is compressed by a booster pump and stored in a CO_2 storage vessel. A pre-determined and metered amount of CO_2 is introduced into the reactor vessel from the bottom

(**Figure 1**). Prior to the reactive treatment, the reactor vessel loaded with the catalyst needs to be purged and rinsed with gaseous CO_2 three times to expel any of the residual moisture and air. Then, the system is gradually pressurized to a pre-determined pressure condition, according to the P-V-T values estimated using the Peng-Robinson equation of state. Different temperatures, pressures, and treatment times have been compared and optimized to find the most effective and beneficial combinations for this reactive treatment step.

After treatment, X-ray powder diffraction (XRD) result clearly showed the transformation from ZnO to $ZnCO_3$, without affecting other catalyst ingredients. The existence of $ZnCO_3$ was reported to be beneficial to the methanol synthesis in various aspects [5,6,7,8]. The crystallite size of copper was also found to be decreased by about 30 percent after the supercritical CO_2 treatment. It has been established that the methanol catalyst activity is quite strongly correlated with the copper crystallite size [5]. At the same time, the cumulative pore volume of the treated catalyst increases about 20 percent on an average, while the internal surface area of the treated catalyst increases about 40 percent.

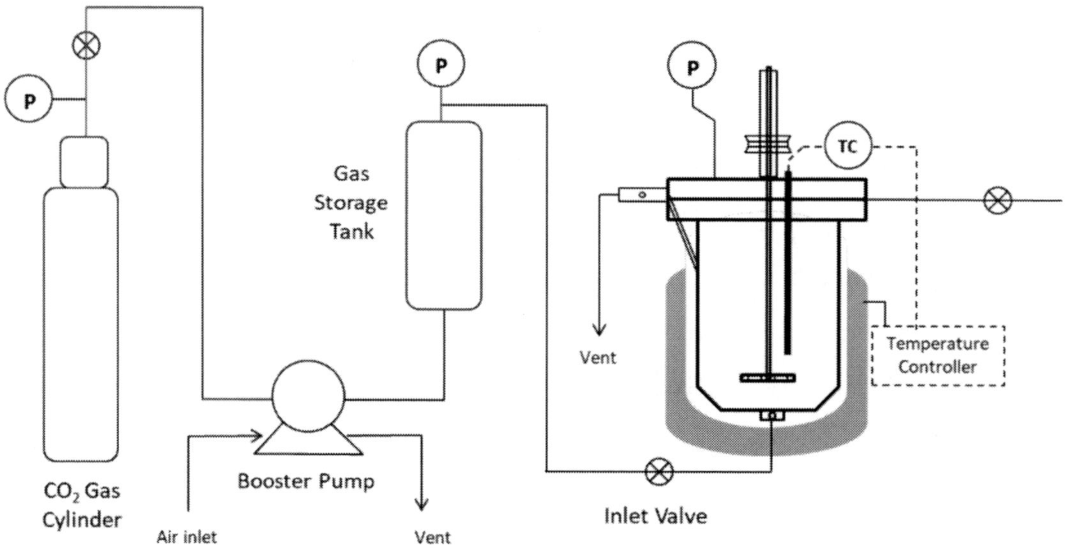

Figure 1 A process flow diagram of CO_2 supercritical treatment and leaching test system

The leaching test was conducted in order to elucidate the detrimental leaching phenomenon in the slurry reactor condition as well as to compare the potential metal leaching severity between the modified catalyst and the original unmodified counterpart. The experiment was carried out in the same reactor vessel with its Magnedrive tuned on (**Figure 1**). For each leaching test, 1.5g of dried catalyst fines of -140 mesh, and 50 ml leaching solution were used. In order to closely simulate the actual reactor conditions expected in the liquid phase methanol synthesis process, the leaching solution for the test was prepared comprising of 45ml SEMTOL 70 oil, 2.5ml distilled water and 2.5 ml pure methanol [7]. The leaching solution with the catalyst slurried was heated to 250°C, and kept under 1000psi pressure with the Magnedrive set at 500rpm. After 8 hrs of slurrying, the reactor was cooled by the built-in tap water circulation and emptied slowly. The catalyst was recovered by separating catalyst fines from the leaching solvent through Whatman-50 filter paper. In order to quantitatively analyze the presence of copper, zinc and aluminum in the leaching solvent, the leaching solvent was ashed using magnesium nitrate as an ashing agent. After ashing, the residue was collected and dissolved in 0.5M HNO_3 aqueous solution for the atomic absorption spectroscopy (AAS) analysis. The detailed ashing procedure and AAS solution preparation procedure are illustrated in **Figure 2**.

The AAS analysis result showed the efficacy of the current modification method on the catalyst. A significantly lower leaching of minerals was detected for the high-pressure CO_2 modified catalyst, e.g., 0.095 mg/L compared with 0.221 mg/L for the non-modified catalyst as for the Zn element. Even though the reactive modification was mainly performed on the ZnO molecules, it also affected the Cu and Al as well. The modified catalyst resulted in 73% lower leaching for copper and 60% lower for aluminum. Such alleviated mineral leaching is significant in preserving the catalyst and its integrity in a harsh environment of a mechanically agitated slurry reactor (MASR) condition. The rationale behind this stabilization is believed to be: (a) steric stabilization due to the presence of larger molecular zinc carbonate in the catalyst matrix and (b) enhanced hydrophobicity of the catalyst internal surface due to the carbonate presence, thus alleviating the interactive effects stemming from water and methanol on the interface.

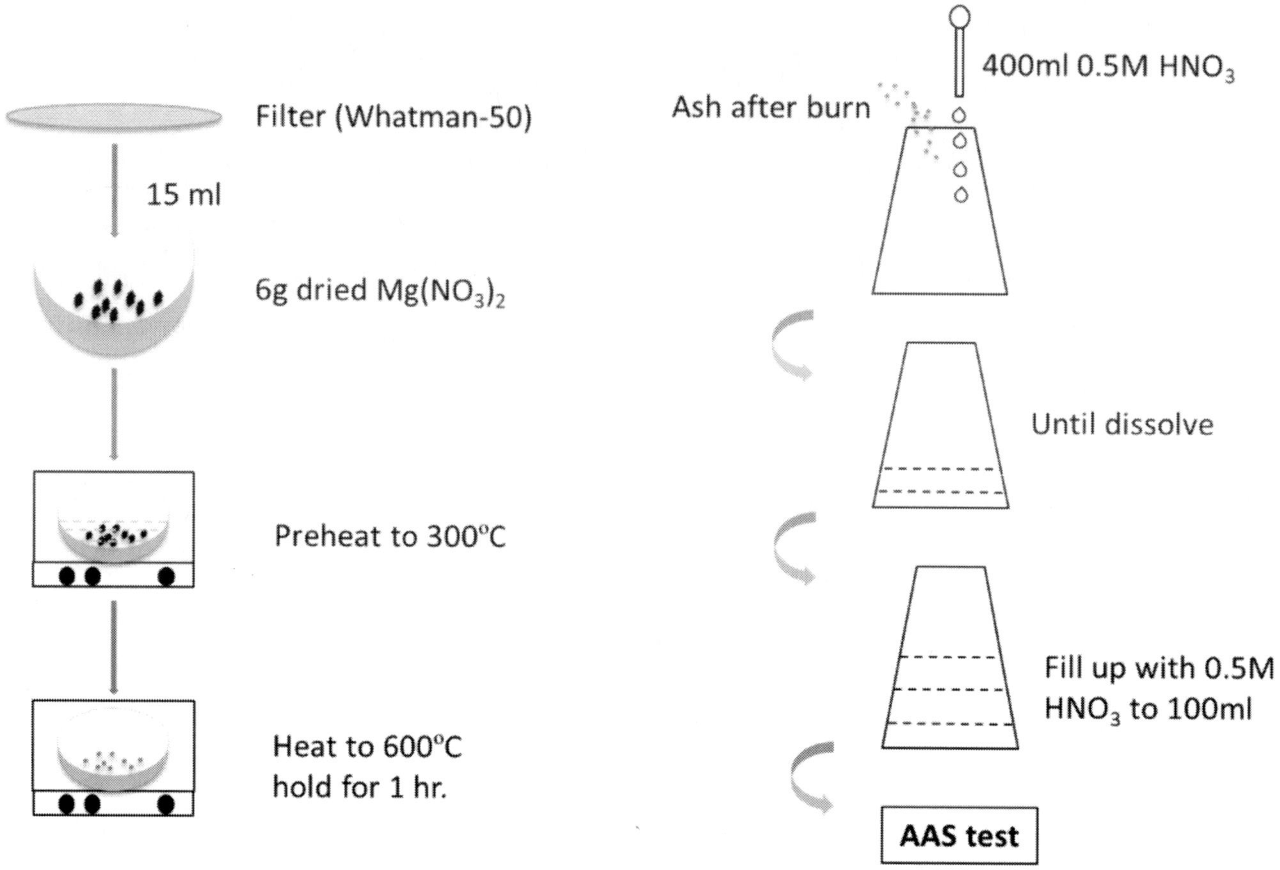

Figure 2. Ashing procedure of leaching solvent and AAS solution preparation for AAS test.

Acknowledgment

This research work was partially supported through research contracts from the Korea Institute of Science and Technology (KIST). The authors are grateful for their support.

References

[1] S. Lee, A. Sawant, K. Rodrigues, and C. J. Kulik, "Effects of carbon dioxide and water on the methanol synthesis catalyst," *Energy Fuels,* vol. 3, no. 1, pp. 2-7, 1989.

[2] M. R. Gogate, A. Foos, S. Lee, and C. J. Kulik, "A single-stage, liquid-phase dinethyl ether synthesis process from syngas III. dual catalyst crystal growth, deactivation and activity conservation studies," *Fuel Science and Technology Int'l,* vol. 8, no. 9, pp. 949-975, 1991.

[3] A. Sawant, M. K. Ko, V. Parameswaran, S. Lee, and C. J. Kulik, "In-situ reduction of a methanol synthesis catalyst in a three-phase slurry reactor," *Fuel Science and Technology Int'l,* vol. 1, no. 5, pp. 77-88, 1987.

[4] S. Lee, B.G. Lee, and C. J. Kulik, "Post-reduction treatment of liquid phase methanol synthesis catalyst using carbon dioxide," *Fuel Science and Technology Int'l,* vol. 8, no. 9, pp. 977-998, 1991.

[5] V. Parameswaran, S. Lee, and I. Wender, "Roles of carbon dioxide in methanol synthesis," *Fuel Science and Technology International,* vol. 7, no. 8, pp. 1021-1057, 1989.

[6] V. Parameswaran, S. Lee, and I. Wender, "The role of water in methanol synthesis," *Fuel Science and Technology International,* vol. 7, no. 7, pp. 899-918, 1989.

[7] S. Lee, Methanol Synthesis Technology, Boca Raton, FL: CRC Press, 1990.

[8] S. Lee, "Process for in-situ regeneration of aged methanol catalysts". U.S. Patent 5004717, 2 April 1991.

Metal Oxide Based Solar-Driven Two-Step CH_4 Reforming and H_2O/CO_2 Splitting Cycle

(Principal Author)
Rahul Bhosale
Department of Chemical Engineering
College of Engineering
Qatar University, Doha, Qatar
rahul.bhosale@qu.edu.qa

Prepared for Presentation at
American Institute of Chemical Engineers
2018 Spring Meeting
14th Global Congress on Process Safety
Orlando, Florida
April 22-26, 2018

Metal Oxide Based Solar-Driven Two-Step CH_4 Reforming and H_2O/CO_2 Splitting Cycle

Rahul Bhosale
Department of Chemical Engineering
College of Engineering
Qatar University, Doha, Qatar
rahul.bhosale@qu.edu.qa

Keywords: Methane reforming, thermodynamics, solar energy, syngas, hydrogen

Abstract

The thermodynamic equilibrium analysis of the solar-driven Tb_2O_3-based CH_4 reforming process is performed. The equilibrium analysis indicates that with the rise in the CH_4/Al_2O_3 molar ratio, formation of Tb and syngas (via methanothermal reduction of Tb_2O_3) enhances and reaches its maximum at 3100 K in case of CH_4/Tb_2O_3 molar ratio = 3.

1. Introduction

Syngas is the primary precursor for the production liquid transportation fuel. In recent investigations, attempts were made to production syngas via H_2O/CO_2 splitting using concentrated solar energy. For instance, various metal oxide based thermochemical H_2O/CO_2 splitting cycles such as iron oxide cycle[1–5], zinc/zinc oxide cycle[6–9], tin/tin oxide cycle[10–12], ferrite cycle[13–21], ceria cycle[22–28], perovskite cycles[29–35] are investigated for the production of H2 and CO which can be combined to produce solar syngas. Although these cycles are successful towards the production of H_2 and CO, the temperatures needed (mainly to reduce the metal oxide thermally) are considerably high. It is possible to reduce the required thermal reduction temperatures of the metal oxides by using methane as the reducing agent (methanothermal reduction of metal oxide)[36]. Due to the inclusion of methane, the syngas can be produced in both thermal reduction and H_2O/CO_2 splitting steps. The redox reactions associated with the metal oxide based methane reforming and H_2O/CO_2 splitting cycle are presented below:

$$MO + CH_4 \rightarrow MO_{red} + H_2 + CO \quad (1)$$

$$MO_{red} + H_2O + CO_2 \rightarrow MO + H_2(g) + CO(g) \quad (2)$$

In this paper, the equilibrium thermodynamic composition associated with the terbium oxide based methane reforming and H_2O/CO_2 splitting cycle are identified using HSC Chemistry 7.1.

2. Equilibrium Analysis

The variation in the equilibrium composition associated with the direct thermal decomposition of Tb_2O_3 in the absence of inert Ar or CH_4 is presented in Fig. 1. The thermal reduction of Tb_2O_3 starts at around 4200 K and the complete dissociation can be achieved at 4480 K. This analysis indicated that to achieve total thermal reduction of Tb_2O_3 (in the absence of CH_4), extremely high temperatures are needed.

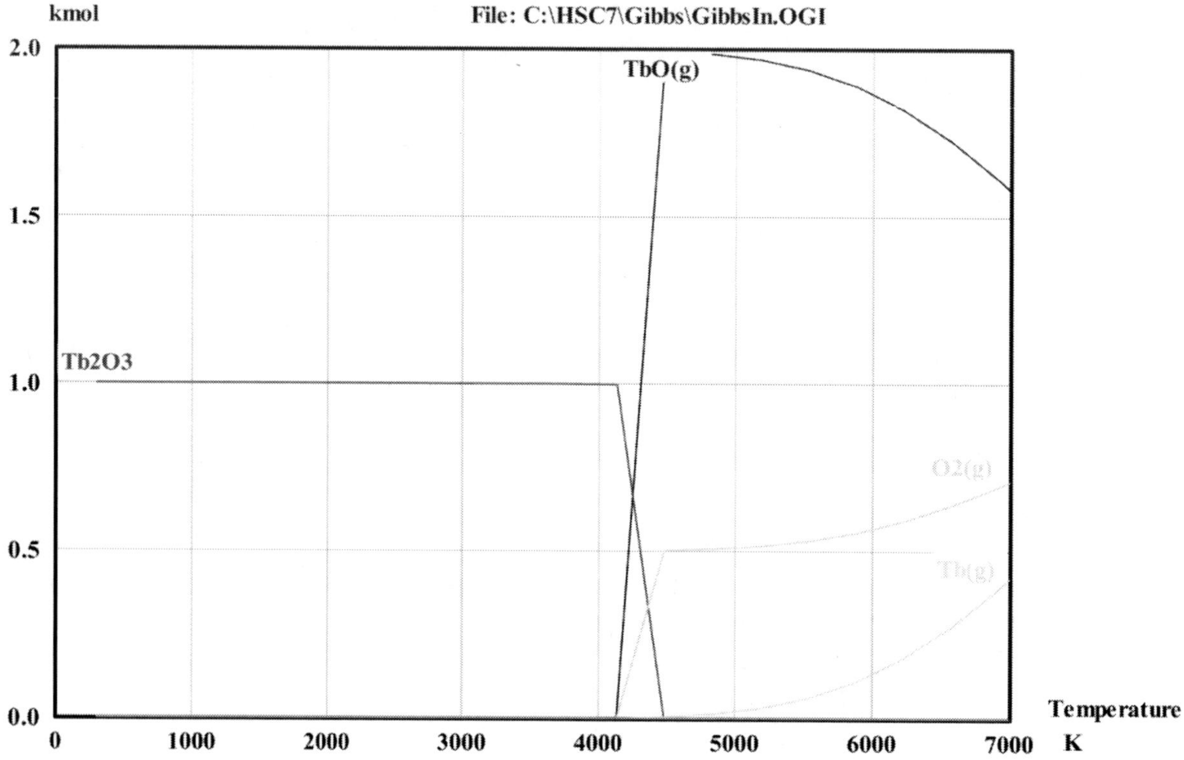

Figure 1. Thermal decomposition of Tb_2O_3 in the absence of inert Ar or CH_4.

By varying the molar flow rate of CH_4/Al_2O_3 in the range of 0.25 to 3 (by keeping $Tb_2O_3 = 1$ mol), the effect of CH_4 as a reducing agent on thermal decomposition of Tb_2O_3 is studied. As shown in Fig. 2 (all sections), as the CH_4/Al_2O_3 molar ratio increases, the thermal reduction temperature (T_H) in case of Tb_2O_3 reduces significantly.

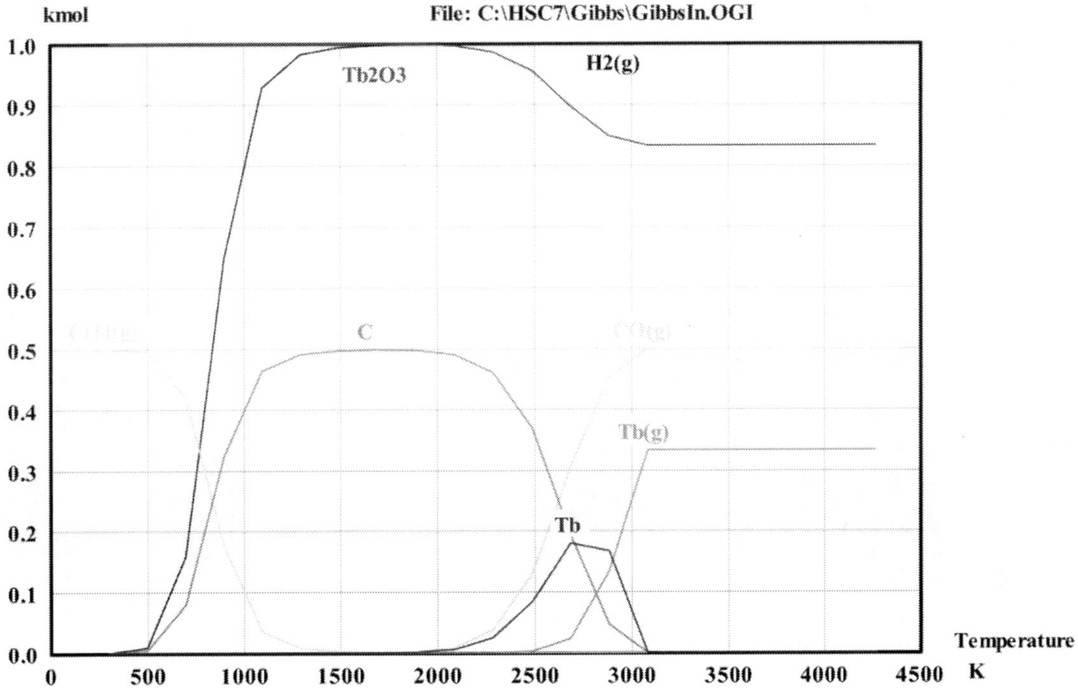

Figure 2. Thermal decomposition of Tb_2O_3 (CH_4/ Tb_2O_3 molar ratio = 0.5).

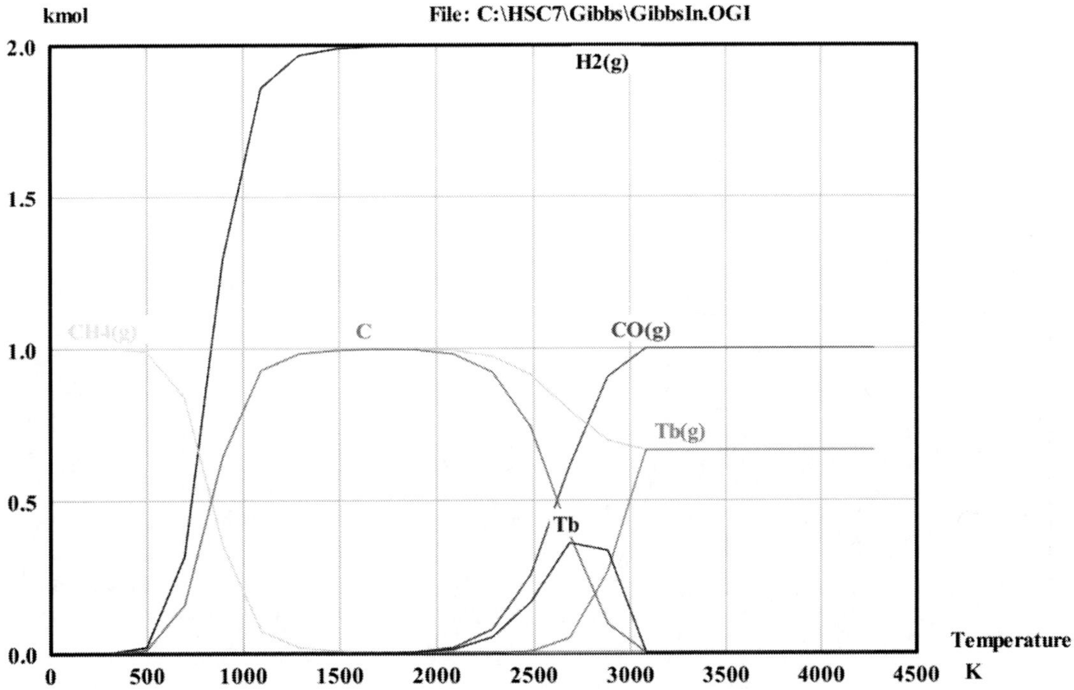

Figure 3. Thermal decomposition of Tb_2O_3 (CH_4/ Tb_2O_3 molar ratio = 1.0).

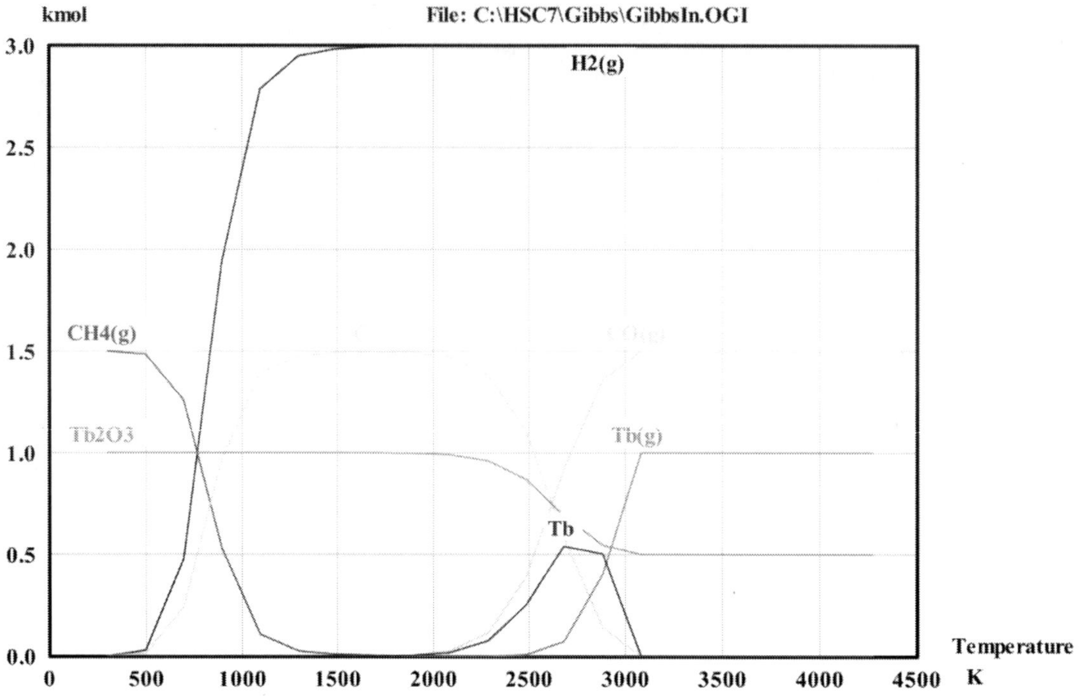

Figure 4. Thermal decomposition of Tb_2O_3 (CH_4/ Tb_2O_3 molar ratio = 1.5).

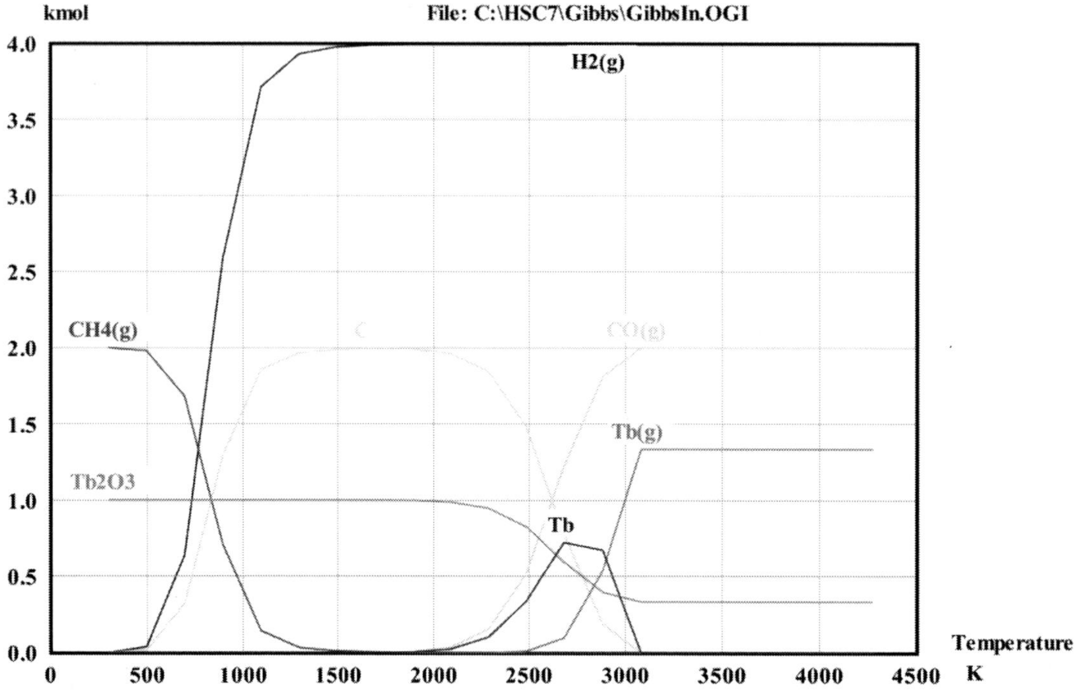

Figure 5. Thermal decomposition of Tb_2O_3 (CH_4/ Tb_2O_3 molar ratio = 2.0).

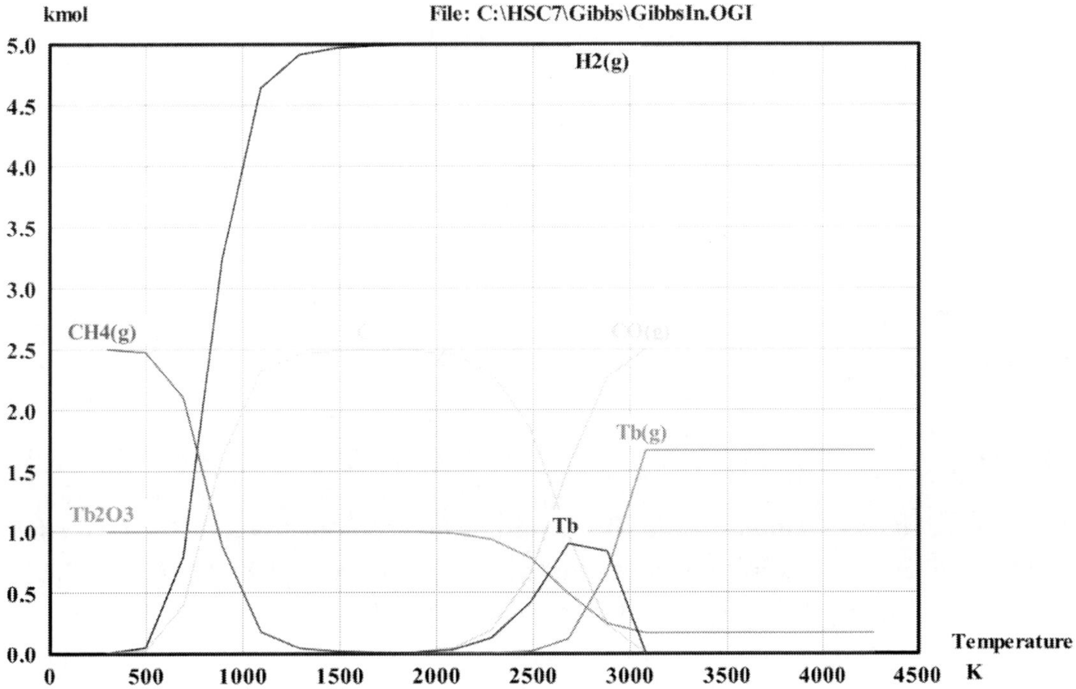

Figure 6. Thermal decomposition of Tb_2O_3 (CH_4/ Tb_2O_3 molar ratio = 2.5).

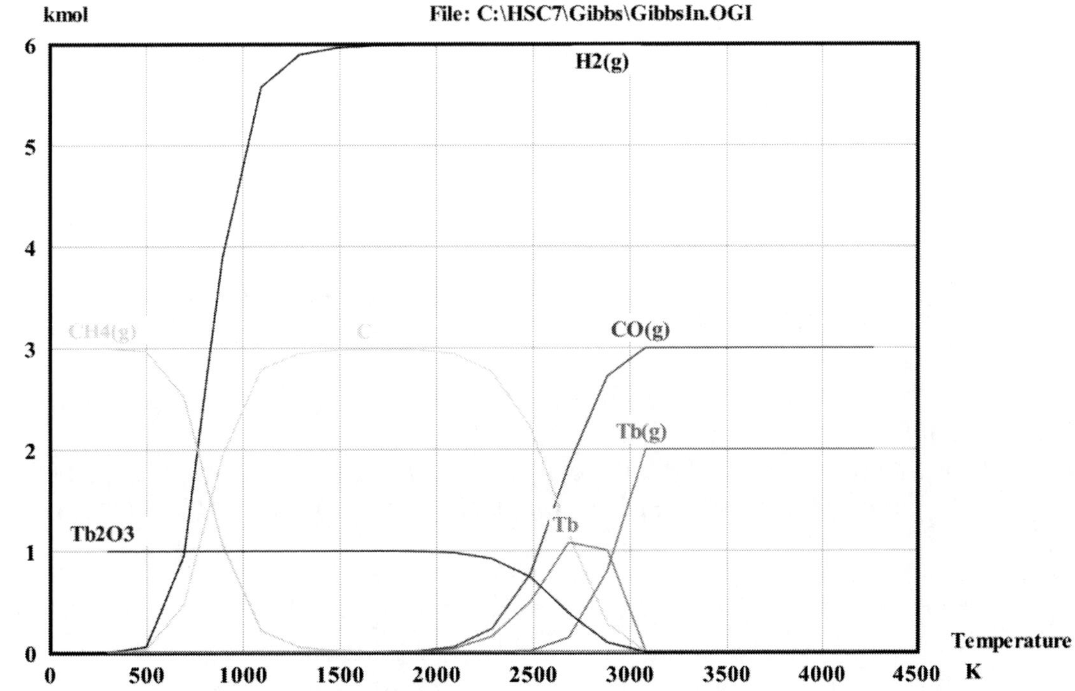

Figure 7. Thermal decomposition of Tb_2O_3 (CH_4/ Tb_2O_3 molar ratio = 3.0).

At all CH_4/Al_2O_3 molar ratios, the thermal reduction of Tb_2O_3 starts at around 2000 K. By keeping the T_H constant (3100 K), the thermal reduction of Tb_2O_3 can be achieved up to 16.5%, 33%, 50%, 67%, 85%, and 100% by varying the CH_4/Tb_2O_3 molar ratio as 0.5, 1.0, 1.5, 2.0, 2.5, and 3.0. At all CH_4/Tb_2O_3 molar ratios, the H_2 production commences at around 400 K. The molar concentration of H_2 upsurges from 1 to 6 moles as the CH_4/Tb_2O_3 molar ratio rises from 0 to 3. The maxima in case of H_2 is attained in the temperature range of 1600 to 1700 K. Further increase in the T_H (at constant CH_4/Al_2O_3 molar ratio) does not affect the H_2 production. Similar to H_2, C formation also starts at 400 K (at all CH_4/Al_2O_3 molar ratio). The carbon formation decreases to 0 moles when the thermal reduction is carried out at 3100K. The CO generation starts at 2000 K and maxima can be achieved at around 3100 K.

4. Conclusion

Effect of CH_4/Tb_2O_3 molar ratio on equilibrium compositions associated with the methanothermal reduction of Tb_2O_3 is investigated in detail. For instance, when CH_4/Tb_2O_3 molar ratio = 1, the partial thermal reduction of Tb_2O_3 can be achieved (33.0%) at 3100 K. However, as the CH_4/Al_2O_3 molar ratio increases up to 2 and 3, the % thermal reduction of Tb_2O_3 upsurges up to 67.0% and 100% (at 3100 K).

5. Acknowledgments

This publication was made possible by the NPRP grant (NPRP8-370-2-154) from the Qatar National Research Fund (a member of Qatar Foundation). The statements made herein are solely the responsibility of author(s).

References

(1) Agrafiotis, C. C.; Pagkoura, C.; Lorentzou, S.; Kostoglou, M.; Konstandopoulos, A. G. *Catal. Today* **2007**, *127* (1–4), 265–277.
(2) Kodama, T.; Nakamuro, Y.; Mizuno, T. *J. Sol. Energy Eng.* **2006**, *128* (1), 3.
(3) Gokon, N.; Murayama, H.; Umeda, J.; Hatamachi, T.; Kodama, T. *Int. J. Hydrogen Energy* **2009**, *34* (3), 1208–1217.
(4) Miller, J. E.; Allendorf, M. D.; Diver, R. B.; Evans, L. R.; Siegel, N. P.; Stuecker, J. N. *J. Mater. Sci.* **2008**, *43* (14), 4714–4728.
(5) Bhosale, R. R.; Kumar, A.; Van Den Broeke, L. J. P.; Gharbia, S.; Dardor, D.; Jilani, M.; Folady, J.; Al-Fakih, M. S.; Tarsad, M. A. *Int. J. Hydrogen Energy* **2015**, *40* (4), 1639–1650.
(6) Steinfeld, A. *Int. J. Hydrogen Energy* **2002**, *27* (6), 611–619.
(7) Abanades, S. In *Industrial and Engineering Chemistry Research*; 2012; Vol. 51, pp 741–750.
(8) Dardor, D.; Bhosale, R.; Gharbia, S.; Kumar, A.; AlMomani, F. *J. Emerg. Trends Eng. Appl. Sci.* **2015**, *6* (September 2016), 129–135.
(9) Koepf, E.; Villasmil, W.; Meier, A. *Appl. Energy* **2016**, *165*, 1004–1023.
(10) Abanades, S.; Charvin, P.; Lemont, F.; Flamant, G. *Int. J. Hydrogen Energy* **2008**, *33*

(21), 6021–6030.
(11) Charvin, P.; Abanades, S.; Lemont, F.; Flamant, G. *AIChE J.* **2008**, *54* (10), 2759–2767.
(12) Bhosale, R. R.; Kumar, A.; Sutar, P. *Energy Convers. Manag.* **2017**, *135*, 226–235.
(13) Lorentzou, S.; Agrafiotis, C. C.; Konstandopoulos, A. G. *Granul. Matter* **2008**, *10* (2), 113–122.
(14) Scheffe, J. R.; Li, J.; Weimer, A. W. *Int. J. Hydrogen Energy* **2010**, *35* (8), 3333–3340.
(15) Muhich, C. L.; Aston, V. J.; Trottier, R. M.; Weimer, A. W.; Musgrave, C. B. *Chem. Mater.* **2016**, *28* (1), 214–226.
(16) Muhich, C. L.; Evanko, B. W.; Weston, K. C.; Lichty, P.; Liang, X.; Martinek, J.; Musgrave, C. B.; Weimer, A. W. *Science (80-.).* **2013**, *341* (6145), 540–542.
(17) Bhosale, R. R.; Shende, R. V.; Puszynski, J. A. *MRS Proc.* **2012**, *1387*, mrsf11-1387-e09-07.
(18) Bhosale, R. R.; Kumar, A.; AlMomani, F.; Ghosh, U.; Sutar, P.; Takalkar, G.; Ashok, A.; Alxneit, I. *Ceram. Int.* **2017**, *43* (6), 5150–5155.
(19) Bhosale, R. R.; Shende, R. V.; Puszynski, J. A. *Int. J. Hydrogen Energy* **2012**, *37* (3), 2924–2934.
(20) Bhosale, R. R.; Alxneit, I.; van den Broeke, L. L. P.; Kumar, A.; Jilani, M.; Gharbia, S. S.; Folady, J.; Dardor, D. Z. *MRS Proc.* **2014**, *1675*, mrss14-1675-rr06-10.
(21) Bhosale, R. R.; Kumar, A.; Almomani, F.; Alxneit, I. *Ceram. Int.* **2016**, *42* (2), 2431–2438.
(22) Abanades, S.; Flamant, G. *Sol. Energy* **2006**, *80* (12), 1611–1623.
(23) Ackermann, S.; Scheffe, J. R.; Steinfeld, A. *J. Phys. Chem. C* **2014**, *118* (10), 5216–5225.
(24) Bhosale, R. R.; Kumar, A.; Almomani, F.; Ghosh, U.; Al-Muhtaseb, S.; Gupta, R.; Alxneit, I. *Ceram. Int.* **2016**, *42* (8), 9354–9362.
(25) Lin, F.; Rothensteiner, M.; Alxneit, I.; van Bokhoven, J. A.; Wokaun, A. *Energy Environ. Sci.* **2016**, 2400–2409.
(26) Bhosale, R. R.; Kumar, A.; AlMomani, F.; Alxneit, I. *Ceramics International*. 2015.
(27) Ackermann, S.; Sauvin, L.; Castiglioni, R.; Rupp, J. L. M.; Scheffe, J. R.; Steinfeld, A. *J. Phys. Chem. C* **2015**, *119* (29), 16452–16461.
(28) Chueh, W. C.; Abbott, M.; Scipio, D.; Haile, S. M. *Science (80-.).* **2010**, *63* (December), 2010.
(29) Scheffe, J. R.; Weibel, D.; Steinfeld, A. In *Energy and Fuels*; 2013; Vol. 27, pp 4250–4257.
(30) Nair, M. M.; Abanades, S. *ChemistrySelect* **2016**, *1* (15), 4449–4457.
(31) Carrillo, R. J.; Scheffe, J. R. *Solar Energy*. 2017.
(32) Dey, S.; Rao, C. N. R. *ACS Energy Lett.* **2016**, *1* (1), 237–243.
(33) Dey, S.; Naidu, B. S.; Govindaraj, A.; Rao, C. N. R. *Phys. Chem. Chem. Phys.* **2014**, *17* (1), 122–125.
(34) Takacs, M.; Hoes, M.; Caduff, M.; Cooper, T.; Scheffe, J. R.; Steinfeld, A. *Acta Mater.* **2016**, *103*, 700–710.
(35) Rahul R. Bhosale, Anand Kumar, Anchu Ashok, Parag Sutar, Gorakshnath Takalkar, Majeda Khraisheh, Fares AlMomani, U. G. *MRS Adv.* **2017**, No. May, 1–8.
(36) Bhosale, R. R.; Kumar, A.; Almomani, F.; Ghosh, U.; Dardor, D.; Bouabidi, Z.; Ali, M.; Yousefi, S.; Alnouss, A.; Anis, M. S.; Usmani, M. H.; Ali, M. H.; Azzam, R. S.; Banu, A. *Energy Convers. Manag.* **2016**, *112*, 413–422.

513635 Biogas Production and Hydrogen Gas Injection Effect in Expanded Granular Sludge Bed Reactor Under Thermophilic Temperature Range

Abstract:

The biogas produced from the anaerobic digestion process could be utilized generally in all the equipment that use natural gas for operation with a minimum adjustment for the lower BTU content of the biogas. Using the EGSB reactor, which is the updated version of the up-flow anaerobic sludge blanket (UASB) under thermophilic range will accelerate the process and rise the biogas production more than operating under other temperature ranges. Generally, there are three different temperature ranges, psychrophilic, mesophilic, and thermophilic ranges. Increasing the temperature will increase the activity of the microorganisms that reside inside the biomass granular particles until it reaches the optimum. However, increasing the temperature above that would cause instability for the microorganisms and reduce the biogas production rate. Hydrogen gas that produced from the first stage of the anaerobic digestion could be utilized and injected back to the EGSB. The hydrogen consuming methanogenic bacteria would start consuming the hydrogen gas as well as the carbon dioxide produced from the first stage and generate methane gas. In this approach, the biogas produced will be enhanced quantitatively and qualitatively. The amount of the hydrogen gas injection will be lower than the consumption ability of the microorganisms in order not to cause inhibition inside the reactor due to the pH drop.

511127 Low Temperature Electrochemical Methods for the Conversion of Bio-Oils to Fuel Intermediates

Abstract:

Bio-oil is a complex mixture of hundreds of organic and inorganic compounds. The reactive species in bio-oil complicate storage, transportation, and downstream processing because of secondary reactions. As such, these oils require upgrading to be used in the conventional transportation fuel infrastructure. The current design case for upgrading of pyrolysis oils uses a complex three-stage process, where the reactors are operated at high temperatures and up to 2000 psi of hydrogen. In the past we used an electrolysis system to hydrogenate reactive functionalities to more stable compounds using protons and water as the hydrogen source. Starting with untreated bio-oils we were able to demonstrate a reduction in the carbonyl content and the phenolic content in a flow cell. In the current work we studied representative molecules to understand the role competition in ECR. Promising results have been observed where significant hydrogenation has been found. A description of work conducted and discussion of results will be presented.

512241 Effect of Biomass Type, Heating Rate, and Sample Size on Microwave Enhanced Fast Pyrolysis Product Yields and Qualities

Abstract:

Using fast pyrolysis processes, renewable biomass resources can be converted to bio-oils that retain more than 60% of the original energy content, and these oils can then be upgraded to motor fuels or other commodity chemicals. The products from fast pyrolysis of biomass are variable and highly dependent upon biomass composition, particle size and geometry, and operating conditions. To better understand the interconnection of biomass characteristics and fast pyrolysis products, a microwave-enhanced fast pyrolysis (MEFP) reactor was constructed. The use of microwave heating under consistent reaction conditions can help to elucidate the primary effects of the original materialâ€™s composition on the product yield. This work tested the MEFP conversion of fifteen biomass materials and blends, including woody feedstock, with and without bark, agricultural residues, and herbaceous energy crops, including some of the materials at different final temperatures and different heating rates. The highest liquid yield was obtained from tulip poplar (64.9 wt%), while the lowest yield was obtained from corn stover (48.5 wt%). The liquid yields obtained from the reactor were repeatable with an average standard deviation of 1.1 wt% (average 1.9% relative standard deviation). The reactor was able to achieve heating rates up to 200 °C/s, although greater product yield was not observed beyond a threshold of approximately 10 °C/s. Multivariate linear regression indicates that >85% of the convertibility of the different feedstocks can be explained using the measured properties of alkali and alkaline earth metals, volatile matter, and lignin contents. Reasonable qualitative agreement was achieved between MEFP yields and yields from a circulating fluidized bed reactor using the same biomass samples (R^2 between 0.78 and 0.85, depending upon how the results are weighted), suggesting that the MEFP can be used as a screening device to select samples and conditions that will maximize the impact from tests performed in larger, continuous conversion systems. MEFP tests have a further advantage in that the microwave heating, which is internal to the particles, facilitates separating the effects of physical heat transfer from those of chemical reaction kinetics for improved understanding of the fundamental polymer deconstruction mechanisms that produce oils from solid feedstocks.

512578 Co-Production of Ethylene and Hydrogen Via Electrochemical Non-Oxidative Deprotonation of Ethane: A Low-Thermal-Budget and Low-Carbon-Footprint Approach

Abstract:

The shale gas revolution in the United States in recent years has resulted in an oversupply of methane and ethane, the major components of natural gas and natural gas liquids, respectively, and this trend will continue for the foreseeable future. Due largely to these cheap resources, e.g., the current price of ethane is less than 20 cents/gal, the chemical industry has moved manufacturing plants back to the United States, creating thousands of jobs, while enjoying healthy margins compared to international rivals that typically use more expensive oil-based naphtha as feedstock. However, the ethylene production from ethane or naphtha is energy intensive and represents the single most energy consuming process in the chemical industry. To fully exploit the potential of ethane as a feedstock, low-thermal-budget and low-carbon footprint processes are required. We demonstrated, for the first time, co-production of ethylene and hydrogen through an electrochemical non-oxidative deprotonation process (NDP) at 400°C, with an ethylene selectivity close to 100% and a hydrogen generation rate of 0.448 mol cm^{-2} per day. Compared to the commercial ethane steam cracking process, the NDP can achieve a 65% in process energy saving and reduce the carbon footprint by as much as 72% or even more (> 90%) if renewable electricity and heat are used. If the heating value of produced hydrogen is taken into account, the electrochemical deprotonation process actually has a net gain in processing energy.

513247 Synthesis and Electrochemical Properties of Ni-Doped TiO_2 Nanotubes for Enhancing Visible Light Absorption

Abstract:

Experimental results have shown that titanium dioxide is a semiconductor with several applications in photocatalytic processes. Its band gap makes it very interesting in the photoelectrodes manufacturing used in photoelectrochemical cells for hydrogen production, a clean and environmentally friendly fuel. The synthesis of 1D titanium dioxide nanostructures, such as nanotubes, makes possible to produce more efficient photoelectrodes for solar energy to hydrogen conversion. In essence, this is because it increases the charge transport rate, decreasing recombination options. However, its principal constraint is to be mainly sensitive to UV range, which represents less than 8% of solar radiation that reaches earth's surface. One of the alternatives to modifying the TiO2â €™s band gap and improving its photoactivity under visible light irradiation is to dope the nanotubes with transition metals. This option requires fabricating efficient nanostructured photoelectrodes with controlled morphology and specific properties able to offer a suitable surface area for metallic doping. Hence, currently one of the central challenges in photoelectrochemical cells is the construction of nanomaterials with a proper band position for driving the reaction while absorbing energy over the VIS spectrum. This research focuses on the synthesis and characterization of Ni-doped TiO_2 nanotubes for improving its photocatalytic activity in solar energy conversion applications. Initially, titanium dioxide nanotubes (TNTs) with controlled morphology were synthesized by two-step potentiostatic anodization of titanium foil. The anodization was carried out at room temperature in an electrolyte composed of ammonium fluoride, deionized water and ethylene glycol. Consequent thermal annealing of as-prepared TNTs was conducted in the air between 450 °C - 550 °C. Afterwards, the nanotubes were superficially modified by nickel deposition. Morphology and crystalline phase of the samples were carried out by SEM, EDS and XRD analysis before and after nickel deposition. Determining the photoelectrochemical performance of photoelectrodes is based on typical electrochemical characterization techniques. Also, the morphological characterization associated electrochemical behavior analysis were discussed to establish the effect of nickel nanoparticles modification on the TiO2 nanotubes. The methodology proposed in this research allows using other transition metal for nanotube surface modification.

Manganese Oxide/Manganese Sulfate Based Solar Thermochemical Water Splitting Cycle

(Principal Author)
Rahul Bhosale
Department of Chemical Engineering
College of Engineering
Qatar University, Doha, Qatar
rahul.bhosale@qu.edu.qa

Prepared for Presentation at
American Institute of Chemical Engineers
2018 Spring Meeting
14th Global Congress on Process Safety
Orlando, Florida
April 22-26, 2018

Manganese Oxide/Manganese Sulfate Based Solar Thermochemical Water Splitting Cycle

Rahul Bhosale
Department of Chemical Engineering
College of Engineering
Qatar University, Doha, Qatar
rahul.bhosale@qu.edu.qa

Keywords: Manganese Oxide – Manganese Sulfate Cycle, Water Splitting, Hydrogen, Thermodynamic Modeling

Abstract

Solar H_2 production via two-step thermochemical Manganese oxide – Manganese Sulfate (MnO-MnS) water splitting cycle is thermodynamically investigated. In the first step, the $MnSO_4$ is thermally reduced to MnO, O_2, and SO_2. In the second step, the reduced MnO is reoxidized via water splitting reaction which results in H_2 production. By using HSC Chemistry software, the thermodynamic equilibrium compositions associated with the thermal reduction and water splitting reactions are identified. Furthermore, the solar-to-fuel energy conversion efficiency of the cycle is determined by performing solar reactor efficiency analysis.

1. Introduction

Among the various metal oxide based thermochemical water splitting cycles such as iron oxide cycle[1–5], zinc/zinc oxide cycle[6–9], tin/tin oxide cycle[10–12], ferrite cycle[13–21], ceria cycle[22–28], perovskite cycles[29–35], sulfur-iodine cycle[36–39], and hybrid sulfur cycle[40–42], the sulfur-iodine cycle and its variation the hybrid sulfur cycle is more appealing as the required operating temperatures are lower as compared to other thermochemical cycles. In these cycles, the SO_3 decomposition is the most energy demanding step. To achieve a noteworthy extent of conversion, the SO_3 decay needs to be carried out under catalytic conditions. While numerous catalytic materials are scrutinized for this application, merely noble metal catalysts are found to be stable towards the sulfur poisoning. The utilization of the noble metal catalyst yields into the significant increases in the cost of the H_2 production.

This paper proposes the utilization of the 'manganese oxide – manganese sulfate' (MnO-MnS) water splitting cycle for the production of solar H_2 via thermochemical redox reactions. The redox reactions involved in this cycles are represented as follows:

$$MnSO_4 \rightarrow MnO + SO_2(g) + {}^1/_2\, O_2(g) \quad (1)$$

$$MnO + SO_2(g) + H_2O(g) \rightarrow MnSO_4 + H_2(g) \quad (2)$$

In this study, the thermodynamic equilibrium compositions associated with both thermal reduction and water splitting step are identified. Furthermore, the solar reactor efficiency analysis is performed and the obtained findings are reported. The principles of the second law of thermodynamics are applied during the solar reactor efficiency analysis.

2. Equilibrium Chemical Composition

2.1 Thermal Reduction of Manganese Sulfate

The equilibrium compositions associated with the thermal reduction of $MnSO_4$ are presented in Figure 1. As per the HSC equilibrium analysis, the complete thermal reduction of $MnSO_4$ into MnO, $SO_2(g)$ and $O_2(g)$ is feasible at/above 1480K (molar flowrate of Ar = 10 mol/s).

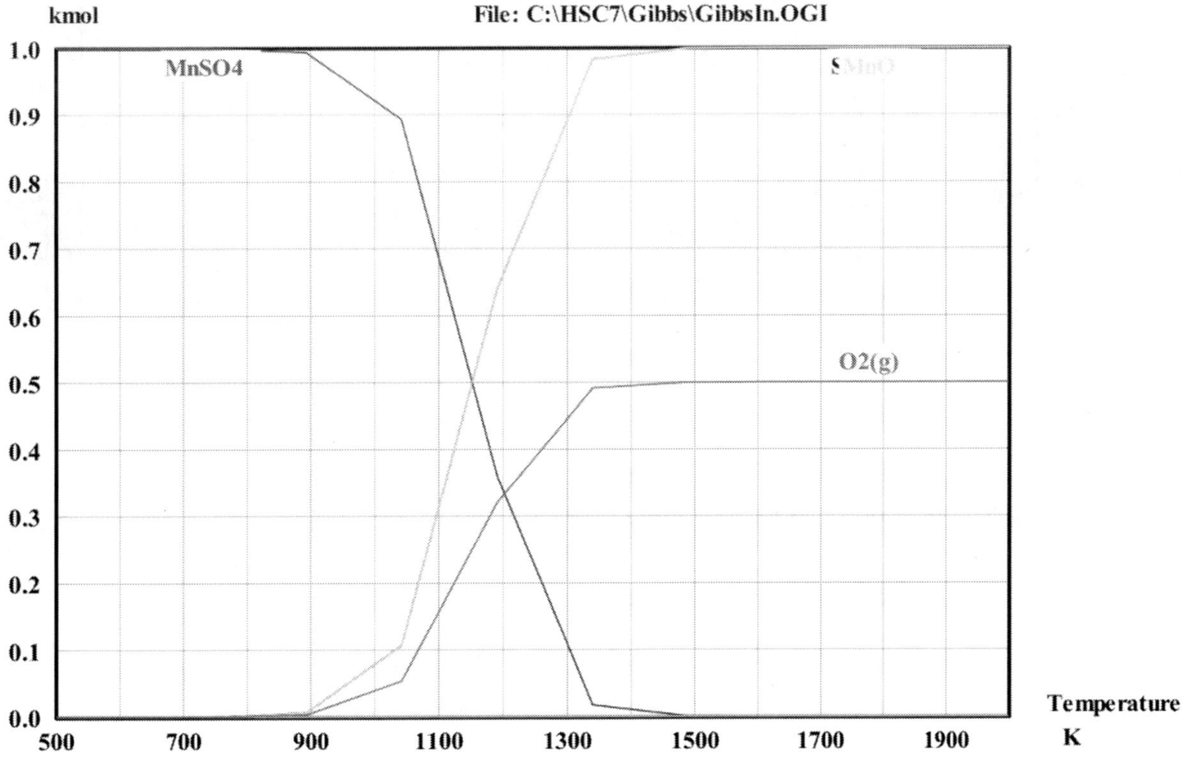

Figure 1. Solar thermal reduction of $MnSO_4$ (Ar = 10 moles/sec).

2.2 Hydrogen Production via H_2O Splitting Reaction

The equilibrium composition related to the water-splitting reaction is presented in Figure 2. According to Figure 2, the H_2 production via re-oxidation of MnO is possible below 540K. The molar concentration of MnO and H_2 are observed to be higher at water splitting temperatures lower than 540K. On the other hand, if the water splitting via MnO re-oxidation is not feasible above 540 K.

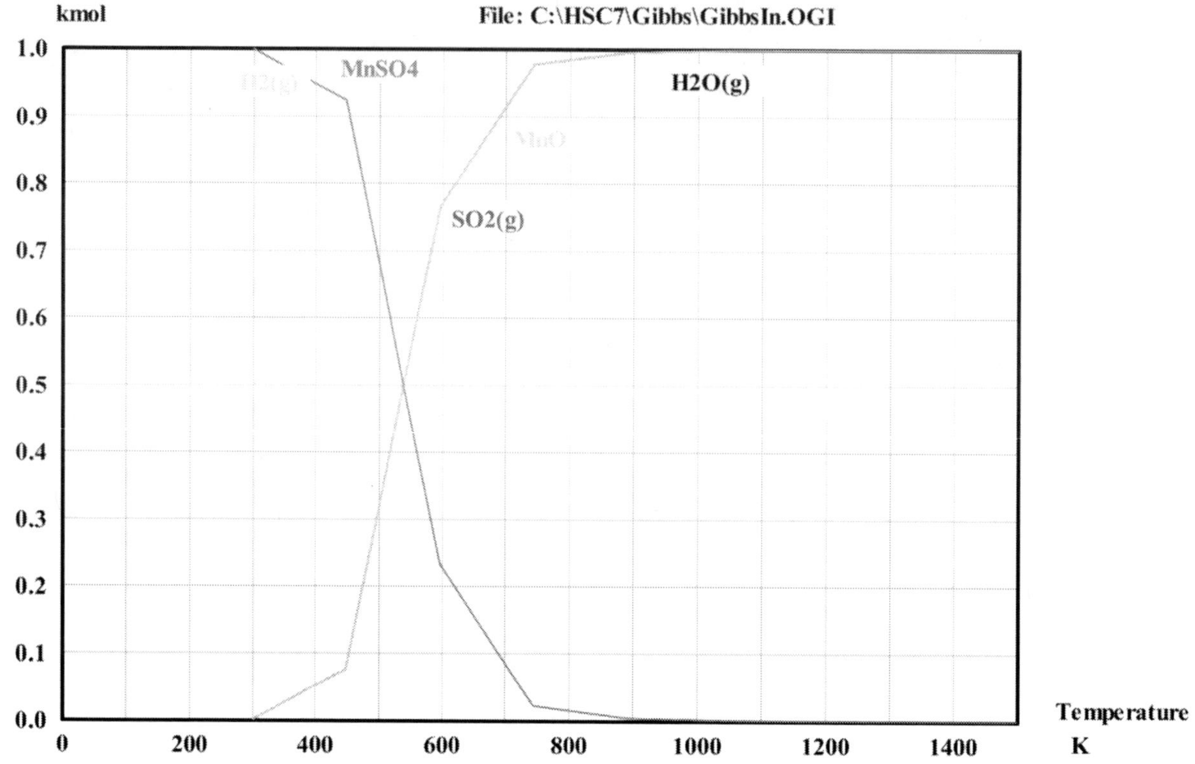

Figure 2. Hydrogen production via water splitting reaction using MnO-MnS cycle.

3. Solar Reactor Efficiency Analysis

The solar-to-fuel energy conversion efficiency of the MnO-MnS water splitting cycle is determined by applying the principles of the second law of thermodynamics. The process flow diagram for the MnO-MnS water splitting cycle is presented in Figure 3. It consists of a solar reactor (to carry out the thermal reduction of $MnSO_4$), water splitting reactor (to reoxidized the reduced MnO and for the production of H_2 via water splitting reaction), an ideal H_2/O_2 ideal fuel cell, a quench unit, and a gas separator. The assumptions employed for the solar reactor efficiency analysis are as follows:

1. The solar reactor is assumed to be a perfectly insulated blackbody absorber
2. H_2 production is carried out at atmospheric pressure and steady state conditions
3. No convective/conductive heat losses
4. Assuming negligible viscous losses and variation in the kinetic and potential energies
5. The energy required for the separation of gaseous products such as separation of SO_2 from O_2 and Ar are not accounted during the efficiency calculation.
6. All reactions reach 100% completion
7. All the products separate naturally without expending any work
8. the heat exchangers required for recovering the sensible latent heat are omitted from thermodynamic considerations
9. Effective emissivity and absorptivity equal to unity

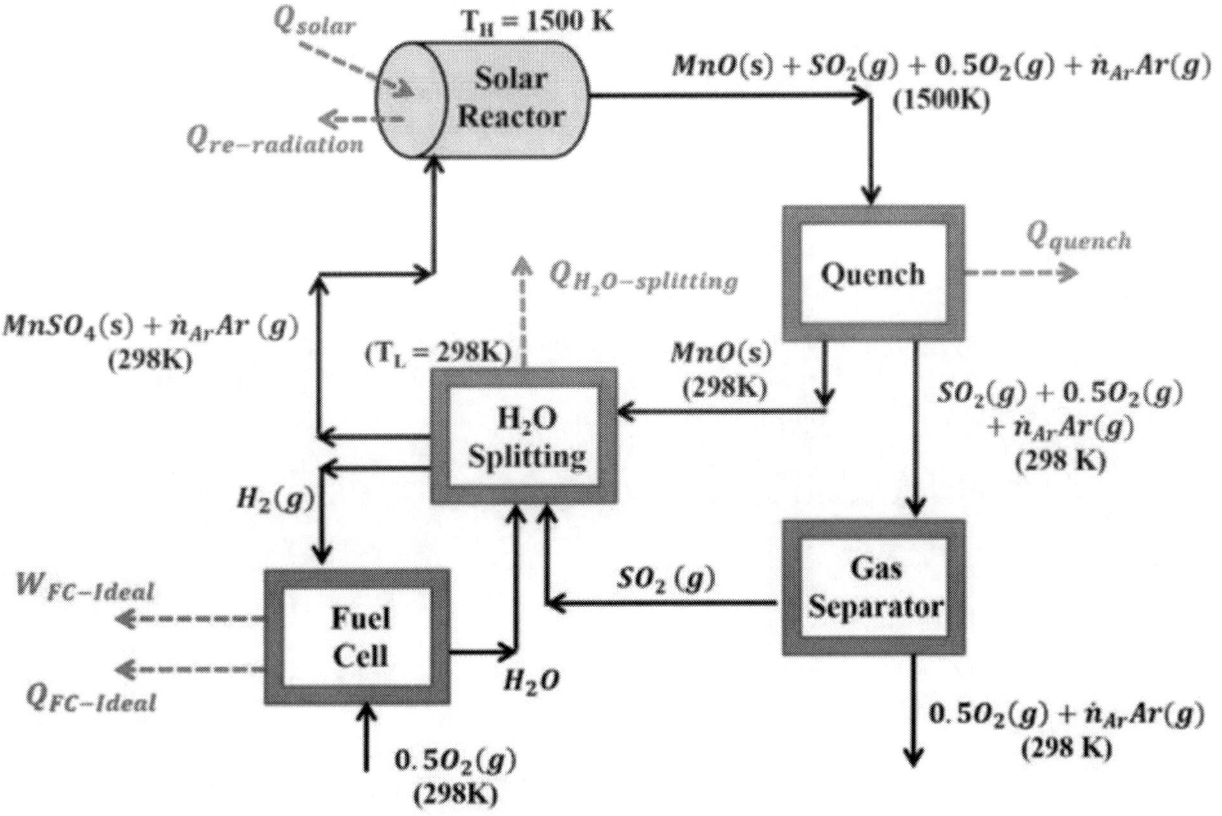

Figure 3. Process flow configuration for H_2 production via two-step solar thermochemical MnO-MnS water splitting cycle.

The methodology applied and the equations derived for the solar reactor efficiency analysis are based on the previous studies reported in the area of thermodynamic analysis of metal oxide based water splitting cycles. Thermodynamic properties are extracted from HSC Chemistry software and databases, and the analysis is normalized to the $MnSO_4$ molar flow rate (1 mol/s) entering the solar reactor.

3.1 Solar energy absorption efficiency ($\eta_{absorption}$)

The solar energy absorption efficiency ($\eta_{absorption}$) is defined as the ratio of the net rate at which solar energy is being absorbed by the solar reactor to the solar power input through the aperture of the solar reactor. The $\eta_{absorption}$ can be calculated as:

$$\eta_{absorption} = 1 - \left(\frac{\sigma T_R^4}{IC}\right) \tag{3}$$

Where, σ = Stefan-Boltzmann constant, I = normal beam insolation $1(kW/m^2)$, C= solar flux concentration ratio of the solar concentrating system (3000 suns), $Q_{reactor-net}$= net energy absorbed in the solar reactor, Q_{solar} = solar energy input. For MnO-MnS water splitting cycle, at T_R= 1500 K, the $\eta_{absorption}$ is observed to be equal to 90.4%.

3.2 *Net solar energy absorbed for the operation of the SO-SS cycle*

The net amount of solar energy needed for the continuous operation of the MnO-MnS water splitting cycle can be calculated by adding the energy required for heating of Ar from ambient conditions up to T_R and energy required for thermal reduction of $MnSO_4$.

$$Q_{reactor-net} = Q_{MnSO_4-reduction} + Q_{Ar-heating} \tag{4}$$

3.2.1 $Q_{Ar-heating}$

In this study, 10 mol/s of inert Ar is used during the complete thermal reduction of $MnSO_4$ at temperatures of 1500 K. According to Eq. (5), 249.9 kW of heat energy is needed to heat the inert Ar from 298 K up to 1500 K (10 mol/s).

$$Q_{Ar-heating} = \dot{n}\Delta H|_{Ar(g)@298K \rightarrow Ar(g)@1500K} \tag{5}$$

3.2.2 $Q_{MnSO_4-reduction}$

The energy required to thermally reduce the $MnSO_4$ into MnO, SO_2 (g), and O_2 (g) is also determined using Eq. (6) and observed to be equal to 530.4 kW.

$$Q_{MnSO_4-reduction} = \dot{n}\Delta H|_{MnSO_4@298K \rightarrow MnO+SO_2(g)+ {}^1/_2 O_2(g)@1500K} \tag{6}$$

By considering both $Q_{Ar-heating}$ and $Q_{MnSO_4-reduction}$, the $Q_{reactor-net}$ for the MnO-MnS cycle is observed to be equal to 780.3 kW.

3.3 *Solar energy input to the solar reactor* (Q_{solar})

By utilizing the estimated values of $Q_{reactor-net}$ and $\eta_{absorption}$, the solar energy input needed to run the MnO-MnS water splitting cycle can be determined as per following equation.

$$Q_{solar} = \frac{Q_{reactor-net}}{\eta_{absorption}} \tag{7}$$

As the $\eta_{absorption} = 90.4\%$, the Q_{solar} value is not very higher as compared to the $Q_{reactor-net}$ and observed to be equal to 863.1 kW.

3.4 *Radiation heat losses from the solar reactor*

Regarding cost and energy saving, it is highly important to determine the heat losses from the solar reactor due to re-radiation. By using the values of Q_{solar} and $Q_{reactor-net}$, the $Q_{re-radiation}$ is calculated as per Eq. (8) and observed to be equal to 82.8 kW.

$$Q_{re-radiation} = Q_{solar} - Q_{reactor-net} \tag{8}$$

3.5 Rate of heat rejected to the surrounding due to quenching

To avoid the recombination of the products, i.e., MnO, $SO_2(g)$, and $O_2(g)$ it is highly essential to quench these products to ambient temperature (298K). To achieve this, a quench unit has been placed in the process, and the amount of heat energy released by the quench unit is calculated as per Eq. (10). The Q_{quench} = 397.0 kW.

$$Q_{quench} = -\dot{n}\Delta H|_{MnO(s)+\ SO_2(g)+0.5O_2(g)+Ar(g)1500K\rightarrow MnO(s)\ +\ SO_2(g)+0.5O_2(g)+Ar(g)@298K} \quad (10)$$

3.6 Rate of heat rejected to the surrounding by MnO oxidizer

After thermal reduction step, the MnO generated is conveyed to the MnO oxidizer, and the H_2 production is achieved via re-oxidation of the MnO (via water splitting reaction). By assuming 100% re-oxidation of MnO via water splitting reaction, the rate of heat rejected to the surrounding by the oxidizer is estimated Eq. (11) and observed to be equal to 141.4 kW

$$Q_{MnO-oxidizer} = -\dot{n}\Delta H|_{MnO+H_2O+SO_2(g)@298K\rightarrow MnSO_4+H_2(g)@298K} \quad (11)$$

3.7 Rate of theoretical work performed and heat rejected by the ideal fuel cell

The ideal fuel cell is placed in the cycle to determine the maximum work that can be extracted from the H_2 produced. The rate of theoretical work performed by the ideal fuel cell is calculated by equation (12) and observed to be equal to 237.05 kW. Likewise, the rate of heat energy released by the ideal fuel cell is determined as equal to 48.56 kW [according to equation (13)].

$$W_{FC-Ideal} = -\dot{n}\ \Delta G|_{H_2(g)+0.5O_2(g)@298K\rightarrow H_2O(l)@298K} \quad (12)$$

$$Q_{FC-Ideal} = -(298) \times \dot{n}\ \Delta S|_{H_2(g)+0.5O_2(g)@298K\rightarrow H_2O(l)@298K} \quad (13)$$

3.8 Solar-to-fuel energy conversion efficiency

The solar-to-fuel energy conversion efficiency of the solar thermochemical MnO-MnS water splitting cycle is defined as the ratio of higher heating value (HHV) of the H_2 produced to the solar energy input:

$$\eta_{solar-to-fuel} = \frac{HHV_{H_2}}{Q_{solar}} \quad (14)$$

Where,

$$HHV_{H_2} = -\dot{n}\ \Delta H|_{H_2(g)+0.5O_2(g)@298K\rightarrow H_2O(l)@298K} \quad (15)$$

For the MnO-MnS water splitting cycle, the $\eta_{solar-to-fuel}$ = 33.1%.

4. Conclusion

The MnO-MnS water splitting cycle has been thermodynamically investigated by using HSC Chemistry software. According to the thermodynamic equilibrium analysis, the complete thermal reduction of $MnSO_4$ (in the presence of Ar = 10 mol/s) is possible at around 1500 K, and the H_2 production via water splitting reaction (with re-oxidation of MnO) is feasible below 540K. The solar reactor efficiency analysis is performed via applying principles of the second law of thermodynamics. The computational thermodynamic analysis indicate $\eta_{absorption}$ = 90.4%, Q_{solar}= 863.1 kW, $Q_{reactor-net}$ = 780.3 kW, $Q_{re-radiation}$ = 82.8 kW, Q_{quench}= 397.0 kW, $Q_{MnO-oxidizer}$ = 141.4 kW, and $\eta_{solar-to-fuel}$= 33.1%, respectively.

5. Acknowledgments

This publication was made possible by the NPRP grant (NPRP8-370-2-154) from the Qatar National Research Fund (a member of Qatar Foundation). The statements made herein are solely the responsibility of author(s).

References

(1) Agrafiotis, C. C.; Pagkoura, C.; Lorentzou, S.; Kostoglou, M.; Konstandopoulos, A. G. *Catal. Today* **2007**, *127* (1–4), 265–277.

(2) Kodama, T.; Nakamuro, Y.; Mizuno, T. *J. Sol. Energy Eng.* **2006**, *128* (1), 3.

(3) Gokon, N.; Murayama, H.; Umeda, J.; Hatamachi, T.; Kodama, T. *Int. J. Hydrogen Energy* **2009**, *34* (3), 1208–1217.

(4) Miller, J. E.; Allendorf, M. D.; Diver, R. B.; Evans, L. R.; Siegel, N. P.; Stuecker, J. N. *J. Mater. Sci.* **2008**, *43* (14), 4714–4728.

(5) Bhosale, R. R.; Kumar, A.; Van Den Broeke, L. J. P.; Gharbia, S.; Dardor, D.; Jilani, M.; Folady, J.; Al-Fakih, M. S.; Tarsad, M. A. *Int. J. Hydrogen Energy* **2015**, *40* (4), 1639–1650.

(6) Steinfeld, A. *Int. J. Hydrogen Energy* **2002**, *27* (6), 611–619.

(7) Abanades, S. In *Industrial and Engineering Chemistry Research*; 2012; Vol. 51, pp 741–750.

(8) Dardor, D.; Bhosale, R.; Gharbia, S.; Kumar, A.; AlMomani, F. *J. Emerg. Trends Eng. Appl. Sci.* **2015**, *6* (September 2016), 129–135.

(9) Koepf, E.; Villasmil, W.; Meier, A. *Appl. Energy* **2016**, *165*, 1004–1023.

(10) Abanades, S.; Charvin, P.; Lemont, F.; Flamant, G. *Int. J. Hydrogen Energy* **2008**, *33* (21), 6021–6030.

(11) Charvin, P.; Abanades, S.; Lemont, F.; Flamant, G. *AIChE J.* **2008**, *54* (10), 2759–2767.

(12) Bhosale, R. R.; Kumar, A.; Sutar, P. *Energy Convers. Manag.* **2017**, *135*, 226–235.

(13) Lorentzou, S.; Agrafiotis, C. C.; Konstandopoulos, A. G. *Granul. Matter* **2008**, *10* (2), 113–122.

(14) Scheffe, J. R.; Li, J.; Weimer, A. W. *Int. J. Hydrogen Energy* **2010**, *35* (8), 3333–3340.

(15) Muhich, C. L.; Aston, V. J.; Trottier, R. M.; Weimer, A. W.; Musgrave, C. B. *Chem. Mater.* **2016**, *28* (1), 214–226.

(16) Muhich, C. L.; Evanko, B. W.; Weston, K. C.; Lichty, P.; Liang, X.; Martinek, J.;

Musgrave, C. B.; Weimer, A. W. *Science (80-.).* **2013**, *341* (6145), 540–542.
(17) Bhosale, R. R.; Shende, R. V.; Puszynski, J. A. *MRS Proc.* **2012**, *1387*, mrsf11-1387-e09-07.
(18) Bhosale, R. R.; Kumar, A.; AlMomani, F.; Ghosh, U.; Sutar, P.; Takalkar, G.; Ashok, A.; Alxneit, I. *Ceram. Int.* **2017**, *43* (6), 5150–5155.
(19) Bhosale, R. R.; Shende, R. V.; Puszynski, J. A. *Int. J. Hydrogen Energy* **2012**, *37* (3), 2924–2934.
(20) Bhosale, R. R.; Alxneit, I.; van den Broeke, L. L. P.; Kumar, A.; Jilani, M.; Gharbia, S. S.; Folady, J.; Dardor, D. Z. *MRS Proc.* **2014**, *1675*, mrss14-1675-rr06-10.
(21) Bhosale, R. R.; Kumar, A.; Almomani, F.; Alxneit, I. *Ceram. Int.* **2016**, *42* (2), 2431–2438.
(22) Abanades, S.; Flamant, G. *Sol. Energy* **2006**, *80* (12), 1611–1623.
(23) Ackermann, S.; Scheffe, J. R.; Steinfeld, A. *J. Phys. Chem. C* **2014**, *118* (10), 5216–5225.
(24) Bhosale, R. R.; Kumar, A.; Almomani, F.; Ghosh, U.; Al-Muhtaseb, S.; Gupta, R.; Alxneit, I. *Ceram. Int.* **2016**, *42* (8), 9354–9362.
(25) Lin, F.; Rothensteiner, M.; Alxneit, I.; van Bokhoven, J. A.; Wokaun, A. *Energy Environ. Sci.* **2016**, 2400–2409.
(26) Bhosale, R. R.; Kumar, A.; AlMomani, F.; Alxneit, I. *Ceramics International*. 2015.
(27) Ackermann, S.; Sauvin, L.; Castiglioni, R.; Rupp, J. L. M.; Scheffe, J. R.; Steinfeld, A. *J. Phys. Chem. C* **2015**, *119* (29), 16452–16461.
(28) Chueh, W. C.; Abbott, M.; Scipio, D.; Haile, S. M. *Science (80-.).* **2010**, *63* (December), 2010.
(29) Scheffe, J. R.; Weibel, D.; Steinfeld, A. In *Energy and Fuels*; 2013; Vol. 27, pp 4250–4257.
(30) Nair, M. M.; Abanades, S. *ChemistrySelect* **2016**, *1* (15), 4449–4457.
(31) Carrillo, R. J.; Scheffe, J. R. *Solar Energy*. 2017.
(32) Dey, S.; Rao, C. N. R. *ACS Energy Lett.* **2016**, *1* (1), 237–243.
(33) Dey, S.; Naidu, B. S.; Govindaraj, A.; Rao, C. N. R. *Phys. Chem. Chem. Phys.* **2014**, *17* (1), 122–125.
(34) Takacs, M.; Hoes, M.; Caduff, M.; Cooper, T.; Scheffe, J. R.; Steinfeld, A. *Acta Mater.* **2016**, *103*, 700–710.
(35) Rahul R. Bhosale, Anand Kumar, Anchu Ashok, Parag Sutar, Gorakshnath Takalkar, Majeda Khraisheh, Fares AlMomani, U. G. *MRS Adv.* **2017**, No. May, 1–8.
(36) Lin, X.; Zhang, Y.; Wang, Z.; Wang, R.; Zhou, J.; Cen, K. *Energy Convers. Manag.* **2014**, *84*, 664–670.
(37) Zhang, Y.; Zhou, J.; Chen, Y.; Wang, Z.; Liu, J.; Cen, K. *Int. J. Hydrogen Energy* **2008**, *33* (20), 5477–5483.
(38) Ginosar, D. M.; Petkovic, L. M.; Burch, K. C. *Int. J. Hydrogen Energy* **2011**, *36* (15), 8908–8914.
(39) Petkovic, L. M.; Ginosar, D. M.; Rollins, H. W.; Burch, K. C.; Deiana, C.; Silva, H. S.; Sardella, M. F.; Granados, D. *Int. J. Hydrogen Energy* **2009**, *34* (9), 4057–4064.
(40) Hinkley, J. T.; O'Brien, J. A.; Fell, C. J.; Lindquist, S. E. *International Journal of Hydrogen Energy*. 2011, pp 11596–11603.
(41) Prosini, P. P.; Cento, C.; Giaconia, A.; Caputo, G.; Sau, S. *Int. J. Hydrogen Energy* **2009**, *34* (3), 1218–1225.
(42) Bayer Botero, N.; Thomey, D.; Guerra Niehoff, A.; Roeb, M.; Sattler, C.; Pitz-Paal, R. *Int.*

J. Hydrogen Energy **2016**, *41* (19), 8008–8019.

509983 Transportation Fuels from Renewable Biomass Sources: Opportunities and Challenges

Abstract:

The production of liquid transportation fuels from biomass in an efficient and cost-effective way is a primary focus of the US DOE. Conversion of wood and other agricultural residual feedstocks through a liquefaction method, such as fast pyrolysis or hydrothermal liquefaction, followed by a catalytic upgrading pathway is one of the most promising ways to produce fuels and chemicals. The resultant oil is a complex liquid mixture of paraffins, aromatics, oxygenates and other components. Main differences between bio-derived oils and petroleum based oils include the high water, acid, and oxygen content of the former. Using current technologies we are able to characterize and account for the majority of the carbon containing species, but still striving to account for 35% unknowns. We also believe that oxygenates present in these oils results in instability and cause complications in down-stream processing. A better knowledge of oxygenates present in bio-oil is needed to properly evaluate bio-oil quality and may provide the key to solving processing issues encountered in the upgrading and refining.

Producing an infrastructure-ready fuel from biomass also requires a better understanding of the chemical constituents of the gasoline, diesel and jet fuel fractions of the biofuel product. Determining the distribution of heteroatoms, especially O and S, impacts the acceptability of a distillate fraction within specific fuel specifications. These reactive and corrosive components can cause complication during downstream processing of bio-oils. We have developed a suite of standardized methods to analyze and describe bio-oils. We refined the analysis of distillates from pyrolysis oil by controlling the source and processing of hydroprocessed bio-oils with two oxygen contents. The upgraded bio-oils were subsequently distilled into five fractions. Analytical techniques such as elemental analysis, density, viscosity, simulated distillation (ASTM D2887), ^{13}C NMR, PIANO and HPLC measurement were used and results will be presented. Specific fractions were found to satisfy boiling point requirements but functional group characterization will likely preclude their direct use as fuel. The methods developed, lessons learnt, and application of these methods to different oils will be discussed.

510481 Thermal Characterization of Microalgal Biomass and Lipids Extracted Residual for Biofuel Production

Abstract:

This communication reports non-isothermal combustion kinetics of microalgal biomass before and after lipid extraction. The thermal decomposition experiments are conducted in a thermogravimetric analyzer (TGA) using air as combustion agent. The biomass samples are exposed up to 800°C with different heating rates. The TGA analysis shows that the decomposition profiles of the biomass samples before and after lipid extraction follow different paths. The apparent activation energy of the oxidation reaction is estimated by fitting an iso-conversional model (Kissingerâ€"Akahira-Sunose) with the experimental data. It is observed that at early stage of biomass combustion, the activation energy increase with increasing the conversion until reaching the maximum at almost a middle conversion stage. After that the value of activation energy decreases sharply. It is also found that the activation energy of lipid extracted biomass decomposition is significantly lower than that of the biomass sample.

Acknowledgments:

Authors like to acknowledge KACST for funding this work through the project NSTIP 13-WAT96-04.

513049 Influence of Carbon Dioxide and Carbonate on the Electrode Reactions in Alkaline Direct Methanol Fuel Cells

Abstract:

Introduction

Direct methanol fuel cells (DMFC) have been among the first fuel cells to be commercialised. The high energy storage capacity and easy handling of the liquid fuels makes the use of DMFC advantageous for applications where a long running time with medium to low power demands is required. Furthermore, the simple and thus robust system design and the typical set-up which keeps the membrane wet for most of the time even after shut-down leads to a high cycle stability. Therefore, DMFC are in particular suitable for applications were random start-up and shut-downs are to be expected like portable power supply, on-board vehicle battery charging or back-up power. Whereas the market for portable fuel cell power systems never really took off outside the military world, battery charging and in particular back-up power are increasing in volume. However, for these markets larger systems in the range of kilowatts are often required. Here the use of today's proton exchange membrane technology based DMFC experiences economical limits due to the need to use platinum based catalyst at rather high loadings. Alkaline anion exchange membrane based DMFC could be an alternative to overcome this issue. However, classic alkaline fuel cells (AFC) are known to be sensitive against carbon dioxide from air. As was shown by Inaba et al. typical anion exchange membranes also suffer from a reduction of conductivity if exposed to carbon dioxide[1]. In contrast to the situation in AFC the effect was, however, found to be reversible. As the current density at which DMFC typically operate are much lower than those of fuel cells operating on hydrogen, the effect of increased ohmic resistance can also be expected to be less relevant.

Much less is known about the effect of CO_2 and carbonate ions on the electrochemical reaction. Vega et al. reported on the effect of CO_2 on the ORR at Pt electrode in KOH electrolyte[2]. A slight reduction of performance was found. For the alcohol oxidation and the oxygen reduction at non PGM catalyst even less is known.

In this study we will report on systematic tests of the methanol oxidation at carbon supported platinum catalyst in KOH and $KHCO_3$ electrolytes as well as the oxygen reduction at commercial PGM-free cathode catalyst K4020 by Acta.. The results will be compared to studies on single cell level. Here tests were performed using also KOH or $KHCO_3$ as supporting electrolyte. Additionally, the effect of dosing CO_2 to the synthetic air feed of the cathode was investigated.

Experimental set-up and procedures

Catalyst tests were performed using a standard rotating disk electrode set-up from PINE instruments with a glassy carbon disk electrode. The disk was coated with an ink of catalyst power dispersed in a mixture of DI water and 2-propanol with a very small amount of added Nafion®. A platinum foil was used as counter electrode and a HydroFlex® reversible hydrogen electrode by Gaskatel as reference electrode. The used cell was equipped with a double jacked so that heating of the cell was possible. Measurements there conducted between 30 °C and 60 °C.

Single cell tests were performed at membrane electrode assemblies consisting of a Fumatech FumaPEM FAA3 electrolyte membrane, an anode GDE made from a Sigracet 10AA GDL coated with Johson & Matthey HiSpec3000 Pt/C catalyst at a platinum loading of 2 mg cm^{-2} and a cathode GDE made from a Freudenberg H32C2 GDL coated with Acta K4020 catalyst at a loading of 5 mg cm^{-2}. Test were performed in a qFC 100-25 cell fixture by balticFuelCells using an Autolab potentiostat as electric load. On the cathode side gas was supplied using mass flow controllers from Bronkhorst for synthetic air and CO_2.

Results

The first observation comparing the methanol oxidation at Pt/C catalyst in 0.25 M KOH and 0.25 M $KHCO_3$ solution is that the achievable peek mass activity in the KOH solution is about a factor of 5 higher than in $KHCO_3$ solution. At the same time the peak potential in the $KHCO_3$ electrolyte is shifted towards lower potentials, whereas the onset potentials are comparable. In both cases a pronounced temperature dependence was observed with the mass activity increasing by about a factor 5 if temperature is raised from 30 °C to 60 °C. A further significant difference between the CV in KOH and $KHCO_3$ electrolyte is the presence of a second oxidation wave in the latter. The CV in $KHCO_3$ resembles thus a CV in acidic electrolyte.

Figure 1: CV of the methanol oxidation in 0.1 M methanol solution at HiSPEC 3000 Pt/C covered electrode as function of temperature left using 0.25 M KOH right 0.25 M $KHCO_3$ as electrolyte.

Chronoamperometric measurement of the methanol oxidation in KOH at different temperatures showed that the initial current is also increasing with temperature. Up to 40 °C the increased temperature does not significantly change the poisoning rate of the electrode. For temperatures of 50 °C and 60 °C an accelerated electrode poisoning of the electrode is observed. In consequence the mass activity at 40 °C exceeds the mass activity at higher temperatures after 10 minutes.

In $KHCO_3$ electrolyte the CA experiments revealed a significant thermal activation of the mass activity only for temperatures of 50 °C and 60 °C. The accelerated degradation here is only observed at 60 °C.

Figure 2: CA measurement of the methanol oxidationm at HiSPEC 3000 Pt/C covered electrodes in 0.1 M MeOH solution at 0.7 V vs. RHE as function of temperature; left in 0.25 M KOH; right in 0.25 M $KHCO_3$.

For the oxygen reduction reaction at Acta K4020 catalyst in KOH electrolyte almost no temperature effect was observed. In $KHCO_3$ at 30 °C the onset potential and half wave potentials are shifted towards lower values by about 200 mV. However, the ORR in $KHCO_3$ exhibited a strong thermal activation so that 60 °C the shift in the onset potential is reduced to about 50 mV.

Figure 3: LSV of the oxygen reduction at Actaka K4020 cobered electrodes, left comparison for 0.25 M KOH and 0.25 M $KHCO_3$ electrolyte at 30 °C; right temperature dependence in 0.25 M $KHCO_3$.

Single cell tests were first performed using 4M methanol, 1 M KOH electrolyte solution as fuel on the anode side. The cathode was supplied with 500 ml min^1 (STP) synthetic air to which CO_2 could be added to yield CO_2 concentrations between 0 ppm and 1600 ppm.

IV curves recorded at 60 °C with 0 ppm and 400 ppm CO_2 addition on the cathode side showed only a small effect of the CO_2 addition The Peak power was reduced by about 2 mW cm^{-2}.

Under galvanostatic conditions of 30 mA cm^2 a stable voltage after an initial voltage decay was observed for all cathodic CO_2 concentrations from 0 ppm to 1600 ppm. With increasing CO_2 concentration on the air side the stable voltage was however slightly reduced.

Figure 4: Single cell tests at 60 °C with 4 ml min^{-1} 4M MeOH, 1M KOH anode feed and 500 ml min^{-1} (STP) synthetic air cathode feed with variable addition of CO_2; left iV curves, right galavnistaic measuremt at 30 mA cm^{-2}

If the electrolyte is changed to 4 M methanol, 2 M $KHCO_3$ solution the cell performance is drastically reduced to values in the 1 mW cm^{-2} range. However, here the addition of 400 ppm to the cathode air supply during iV measurements yielded a raise in the peak performance by about 0.2 mW cm^{-2}. This positive influence of cathodic CO_2 addition on the cell performance with anodic $KHCO_3$ electrolyte could be verified in galvanostatic measurements.

Figure 5: Single cell tests at 60 °C with 4 ml min^{-1} 4M MeOH, M $KHCO_3$ anode feed and 500 ml min^{-1} (STP) synthetic air cathode feed with variable addition of CO_2; left iV curves, right galavnistaic measuremt at 30 mA cm^{-2}

Discussion

Full saturation of the electrolyte with CO_2 leading to $KHCO_3$ electrolyte solution change the electrochemical behaviour of the catalyst significantly. In all cases the obtainable mass activity was reduced. For the methanol oxidation the CV also exhibits typical features of the methanol oxidation in acidic environment. Therefore, the change in pH will play a strong roll. For the methanol oxidation no significant differences in the thermal activation were observed. In contrast to this the ORR in $KHCO_3$ is strongly thermally activated so that at high temperature the differences to KOH become small.

In single cell tests the change from KOH to $KHCO_3$ electrolyte on the anode side had a much stronger effect than the addition of CO_2 on the cathode side. This also indicates the strong influence of the electrolyte on the methanol oxidation kinetics.

1. Inaba M, Matsui Y, Saito M, et al. Effects of carbon dioxide on the performance of anion-exchange membrane fuel cells. *Electrochemistry.* 2011;79(5):322-325.

2. Vega JA, Mustain WE. Effect of CO(2), HCO(3)(-) and CO(3)(-2) on oxygen reduction in anion exchange membrane fuel cells. *Electrochimica Acta.* Feb 1 2010;55(5):1638-1644.

Water Management Strategies for PEM Fuel Cells - a Review

Harshit Bajpai[a], Dr. Shripad T. Revankar[b]
Davidson School of Chemical Engineering, Purdue University, West Lafayette, IN 47907
School of Nuclear Engineering, Purdue University, West Lafayette, IN 47907

Abstract

Water management is an integral part of PEM fuel cell operation. Water has to be adequately balanced inside a fuel cell between hydration, diffusion and electro-osmotic drag processes, as an imbalance can lead to either flooding or drying, which can result in degradation of current density and output voltage. This paper reviews the operating parameters and design of fuel cell components for efficient water balance. It is very crucial to optimize operating parameters viz. pressure, humidity of inlet streams, temperature, flow rates and stoichiometry in a fuel cell. Typically, the optimum values of relative humidity and flow rates are decided by a bargain between proton hydration and water imbalance. The design enhancement strategies can significantly augment the performance of fuel cells. Interdigitated channels are better than conventional parallel flow fields and can double the power density. A microporous layer with hydrophobic substances can improve water diffusion from electrode assembly into the diffusion medium. The higher in-plane permeability of gas diffusion layer leads to a better gas transport and reduces retention volume of water. Thus, fuel cell performance is strongly linked to the adopted strategy.

Introduction

The use of alternative energy sources is growing in importance due to depletion of fossil fuels and environmental concerns. In this scenario, PEM fuel cells (PEMFCs) pose a more efficient and environment friendly substitute of the existing internal combustion engines(ICEs) used for transportation. Unlike ICEs, PEM fuel cells efficiency are not limited by the Carnot cycle and fuel cells can attain higher efficiencies [1]. Demirdöven and Deutch [2] calculated efficiencies using a simple model and showed an advanced fuel cell vehicle had 23.9% efficiency compared to 11.3% of efficiency of combustion engine. In this case, the advanced fuel cell vehicle was considered to have an on board mini refinery to reform gasoline to hydrogen. However, roughly 50 % efficiency can be achieved if the pure hydrogen is supplied as fuel in an 80-kW stack [3]. PEM powered vehicle will also have advantages of zero emissions, reduced noise and simplicity of designing.

PEMFCs typically have a solid per fluorinated polymer electrolyte membrane between two porous electrodes loaded with platinum catalysts to accelerate the oxidation and reduction reactions, this arrangement is also called membrane electrode assembly(MEA). MEA is supported by bipolar plates with porous gas diffusion layers (GDLs) sandwiched between MEA and distributor plates. The GDLs typically consists of a carbon paper treated with hydrophobic substances for water management. The bipolar plates ensure collection of electric current, distribution of reactant gases and removal of produced liquid and gasses. Figure 1 shows a schematic diagram of a PEMFC with insights on the ions and water migration across the cell.

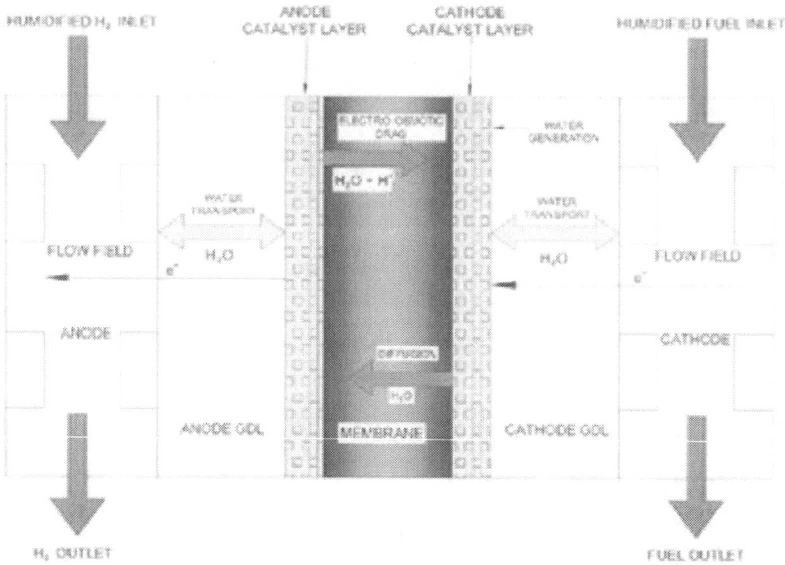

Figure 1: Schematics of a PEM fuel cell

In normal operation, the water is produced at cathode and at anode the produced protons are surrounded by water molecules inside its solvation cell. Generally, a water concentration gradient exists with higher concentration at cathode that results in a diffusion of water molecules from the cathode to anode [4]. However, electric field induced proton movement from anode to cathode drags with it certain water molecules. This phenomenon is known as electro-osmotic drag [4]. The water imbalance can result into two different conditions viz. flooding and drying, the former hinders diffusion of reactants [5] and latter reduces the proton conductivity. Natarajan and Nguyen [6] characterized flooding using segmented cell assembly as a stepwise drop in local current densities which they hypothesized to indicate water filling up the pores of electrodes. Thus, when flooding occurs, less reactants flow through that cell, as a consequence less water is removed leading to a severe starvation condition. Persistent flooding was observed at even low stoichiometry with a two to five-fold increase in pressure drop and a subsequent voltage drop up to 80 mV [7]. Flooding was observed typically at the corner of the channels due to the large contact area which increases the viscous force of water [8]. On the other hand, a dry membrane has 300 times decrease in proton conductivity compared to a fully hydrated membrane [9]. Le Canut et. al [10] noticed that there is a substantial potential drop with decreasing the water content due to higher polarization and ohmic losses. Generally, this drop is temporary which can be recovered by timely humidification, however, long time operation leads to severe irreversible membrane degradations. Pinhole formations wherein membrane thickness is significantly reduced can also result due to dry membrane operation [11].

1. Water management

Water has to be adequately balanced inside a fuel cell between hydration, diffusion and electro-osmotic drag processes, as an imbalance can lead to either flooding or drying, which can result in degradation of current density and output voltage. It is very crucial to optimize operating parameters viz. pressure, humidity of inlet streams, temperature, flow rates and stoichiometry in a fuel cell. Yet, these strategies frequently lead to serious parasitic losses which are associated with pressure drops and flow rates and at times membrane ruptures [12]. The accompanying section describes the effects of these operating parameters on performance.

1.1. System pressure

There exist a direct association of the operating pressure with the fuel cell performance, as efficiency increases with pressure increments [13,14]. The increased pressure drop across the cell, indicates clogging of electrode gas channels due to flooding [15]. Under ideal operating conditions, electro-osmotic drag rate is higher than diffusion rate, which causes cathode over potential. A technique for counterbalancing the rates was developed by Voss et. al. [16] of setting up a pressure gradient ($P_a < P_c$), this will facilitate the movement of produced water from cathode to anode. However, the pressure gradient impacts membrane adversely, so the gradient is limited to about 500 mbar. Pócza et al. [15] suggested that to remove excess water it is essential to setup a pressure drop that results in a driving force higher that the frictional resistance between wall and droplet. Feeding temporarily higher than operating pressure gasses or lowering the exit cell pressure creates higher and lower pressure pulses respectively that aids in removal of water [17]. Therefore, the optimum pressure can help in better performance provided it can be maintained without subsequent cross effects to other aspects of cell.

1.2. System temperature

The interdigitated flow distributors by He et al. [18] showed that the saturation pressure and evaporation rates are increased with increments in operating temperature. Thereby, the excess water inside the cells evaporates, improving the cell performance [19]. Additionally, low surface tension and viscosity of water at high temperatures further aids the flushing of cell. Hence, a flooded cell can be recovered by a sudden temperature increase from 40 to 50 $^{\circ}$C [18]. The fuel cell temperature should not exceed 60 $^{\circ}$C as it can result aggressive drying of membrane due reactants [20,21]. So, temperature can aid in flushing and also improving the performance within a specific temperature range.

1.3. System humidity

In normal operation, humidification of gas is undertaken of inlet gases prior to entry [20]. Conversely, in internal humidification mode the dry gases are humidified when then come in contact with water laden membrane [22]. However, the cell performance is lowered by 40% when operated in internal humidification mode due to less water content or steep water gradients in membrane [21]. There is an increase in current distribution heterogeneities with increase in humidity levels of anode inlet streams due to lesser water diffusion from cathode to anode [23]. Also, there is a subsequent increase in cell voltage with increase in humidity as ohmic overpotential decreases [24]. There is an optimum value of relative humidity for inlet streams, as there is a bargain between proton hydration and cell flooding. The cell can be flushed by dehumidification of one or both inlet streams as water will evaporated from flow fields. Dehumidifying one or both reactant streams is also a corrective action which is used after cell flooding and which would promote water vaporization in the flow fields [19]. Thus, a balance is to be achieved between the high humidification performance and lower energy consumption for economical use of PEMFC.

1.4. System flowrates

As a general practice, the stoichiometric ratios are higher than 2 for air inlet streams. This is done as at lower ratios flooding is favored, and that might cause reactant depletion [20]. Hakenjos et al. [25] noticed that increasing the reactant flowrates, results in higher output current density and flushing of the water in flooded cells. Natarajan and Nguyen [23] used six segmented electrodes assembly with common hydrogen and oxygen entry and exit points. They observed that at a constant voltage the uniformity of current densities is maintained for longer durations at higher oxygen flowrates. They also summarized that if fuel is surplus that current densities area are decided by hydration of MEA, on the other hand, if fuel supply is insufficient, electrochemical reaction dominates, leading to loss in performance due to fuel dearth. Jame [26] showed that operating anode at higher stoichiometric ratios results in higher average current density and increased sensitivity to flooding of cathode. Similar to humidity, the optimum value of air and hydrogen flowrates can be decided by a bargain between removing excess liquid water and drying of membrane.

2. *Cell Design*

The design enhancement strategies can significantly augment the performance of fuel cells and help in mitigating the water management problems. The design augments for flow fields, GDL, MEA, MPL and others are discussed.

2.1. Flow Fields

Effective water management can result, when flow fields can remove water out of the cells that is constantly transported from the GDL. Typically, the fuels cells that are employed with conventional parallel flow fields can only reduce flooding problems by increasing the water evaporation rates [23]. These days, interdigitated designs are used that can additionally lessen mass transport constraints [27]. In these dead-end flow fields, the gasses are forced to flow through GDL, this changes the gasses transports from diffusion driven to forced convection driven mechanism and this force aids in flushing of any water that in other cases collect in the GDL. Nguyen [27] proved experimentally that interdigitated design enhances the performance in comparison with parallel flow fields. Wood et al. [28] also showed that interdigitated flow designs elaborates the operable range to higher current densities and results in a 50 - 100 % increase in performance. This increased performance is correlated to more uniform distribution of reactants across GDL and greater capability to water production. In an experimental comparison of three types of cathode flow channels (parallel, cascade and interdigitated), Liu et al. [29] observed the interdigitated channels are flooded at the last and flooded areas are also reduced. In [30], Nguyen et al., developed a three-cell PEM stack with interdigitated flow fields operating at ambient pressure and temperature on pure hydrogen and oxygen. They showed that use of sequential exhaust control increased peak power from 0.26 W/cm^2 per cell to 0.5 W/cm^2 per cell. Ashrafi and Shams [31] studied the effects of orientation of flow fields and observed that in vertical cells, with channels positioned horizontally and inlet streams are located on the upper side of flow fields recorded lowest pressure drop. This can be attributed to enhanced water removal due to film flow and gravity. The down slanted anode channels were observed to have better performance due to improved proton hydration [32]. This is due to the fact that less water is being collected at cathode of a down slanted anode channel in comparison of a normal rectangular channel as down slanting created a water hydration gradient which caused diffusion of water to the anode.

2.2. GDL and MEA improvements

The cell components should be designed to facilitated water evacuation at cathode. Typically, GDL contains hydrophobic PTFE [33]. The hydrophobic particles can reduce water accumulation performance losses [34] The GDL areas neat reactant inlets are relatively dry than other active areas [24]. Another way to achieve this is by locally increasing cathode porosity near the air outlet, by using some carbon clothes instead of fiber or by drilling holes in carbon fiber in areas of severe water accumulation [35]. Pócza et al. [15] show direct association of use of hydrophobic and hydrophilic substances on carbon papers on water accumulation in gas channels. They showed that in cases of lower temperature of 30 °C hydrophilization of GDL can be very effective as current densities are homogenized. In the case of hydrophobic the discrete droplets of water are able to move due to kinetic energy of reactant flow and decreased surface tension of hydrophobic substances. They attributed higher current densities of hydrophilic papers to improved water content in MEA. Salva et at. [24], observed that cells with GDL having lower in-plane has permeability retained lesser water, this can be explained by in-plane gas permeability dominance over the rate of convective flow through GDL.

2.3. Microporous Sublayer (MPL) effects

The use of MPL typically improves water transport from electrodes to diffusion medium [36]. This is due to a decrease in flooding of catalyst layer and more intimate contact between sublayer and catalyst layer. A PTFE carbon sublayer is added between the carbon paper and the catalyst layer which resulted in significant enhancement of water management [37]. Qi and Kaufman [37] analyzed sublayers containing different percentages of PTFE (24, 35 and 45) with 2 mg/cm2 carbon loading and observed 35% to perform best and 45% performed worst. In a further analysis, PTFE, FEP and PDMS were compared for hydrophobicity nature in MPL [38]. This study showed that PDMS was superior in reducing flooding than PTFE and FEP. In another study, Li and Chan [39] showed the use of hydrophobic DSO- loaded on cathode catalyst layer can have similar effects as of MPL. This DSO loaded layer performance improvement can be attributed to lower cathode diffusion resistance.

2.4. Other advancements

Bae et al. [40] had developed an asymmetric membrane with a combination of hydrophilic and hydrophobic substances to effectively manage water transport. This membrane improved water diffusion from cathode to anode thus increasing the cell performance. The hydrophilic effect was noticed by higher water retention and proton conductivity, on the other hand hydrophobic behavior resulted in better water transport due to repulsion forces. Kratsberg and Ein [41] used a fuel cell in which cathode was impregnated with oxygen permeable water impermeable substances like perflourinated hydrocarbons and they observed enhanced water management. The effect of clamping forces was observed by Cha et al. [42] as ohmic resistance decreased with increase of forces because of higher membrane hydration. Conversely, with increase in forces the resistance increased due to higher water flow resistance.

3. Conclusion

The underlying issue related with performance on PEM fuel cell, water management was reviewed. Its simple associations were understood with various operating conditions and design upgrades. An elementary basis has been laid out regarding the dependency of these parameters on performance and current densities. In general, mitigation of flooding can be achieved by operating at optimum conditions of pressure, temperature, humidity and flow rates. The various design upgrades discussed can certainly aid with proper balance but consequently hinder the commercialization of cells. The identification of the subtle equilibrium between flooding and membrane drying is significant and interdependent on complex multiphase flow dynamics. Thus, extensive research needs to be carried out on visualization of water transport and optimum operational strategies with simple and conveniently reproducible cell designs.

References

1. K. Kordesh & G. Simader, (1996) Fuel Cells and Their Applications, John Wiley & Sons, Ltd, p. 44.

2. Demirdöven, N., & Deutch, J. (2004). Hybrid Cars Now, Fuel Cell Cars Later. Science, 305(5686), 974.

3. DeCicco, J. M. (2004). Fuel Cell Vehicles. In Reference Module in Earth Systems and Environmental Sciences. Elsevier. p 768

4. Zawodzinski, T. A. (1993). Water Uptake by and Transport Through Nafion® 117 Membranes. Journal of The Electrochemical Society, 140(4), 1041.

5. Kramer, D., Zhang, J., Shimoi, R., Lehmann, E., Wokaun, A., Shinohara, K., & Scherer, G. G. (2005). In situ diagnostic of two-phase flow phenomena in polymer electrolyte fuel cells by neutron imaging: Part A. Experimental, data treatment, and quantification. Electrochimica Acta, 50(13), 2603–2614.

6. Natarajan, D., & Van Nguyen, T. (2005). Current distribution in PEM fuel cells. Part 1: Oxygen and fuel flow rate effects. AIChE Journal, 51(9), 2587–2598.

7. Hussaini, I. S., & Wang, C. Y. (2009). Visualization and quantification of cathode channel flooding in PEM fuel cells. Journal of Power Sources, 187(2), 444–451.

8. Su, A., Weng, F. B., Hsu, C. Y., & Chen, Y. M. (2006). Studies on flooding in PEM fuel cell cathode channels. International Journal of Hydrogen Energy, 31(8), 1031–1039.

9. Sone, Y., Ekdunge, P., & Simmonsson, D. (1996). Proton Conductivity of Nafion 117 as Measured by a Four-Electrode AC Impedance Method. Journal of The Electrochemical Society, 143(4), 1254.

10. Su, A., Weng, F. B., Hsu, C. Y., & Chen, Y. M. (2006). Studies on flooding in PEM fuel cell cathode channels. International Journal of Hydrogen Energy, 31(8), 1031–1039.

11. Lee, H., Kim, T., Sim, W., Kim, S., Ahn, B., Lim, T., & Park, K. (2011). Pinhole formation in PEMFC membrane after electrochemical degradation and wet/dry cycling test. Korean Journal of Chemical Engineering, 28(2), 487–491.

12. Li, H., Tang, Y., Wang, Z., Shi, Z., Wu, S., Song, D., … Mazza, A. (2008). A review of water flooding issues in the proton exchange membrane fuel cell. Journal of Power Sources, 178(1), 103–117.

13. Aslam, R. M., Ingham, D. B., Ismail, M. S., Hughes, K. J., Ma, L., & Pourkashanian, M. (2017). Simultaneous direct visualisation of liquid water in the cathode and anode serpentine flow channels of proton exchange membrane (PEM) fuel cells. Journal of the Energy Institute, 1–14.

14. Yousfi-Steiner, N., Moçotéguy, P., Candusso, D., Hissel, D., Hernandez, A., & Aslanides, A. (2008). A review on PEM voltage degradation associated with water management: Impacts, influent factors and characterization. Journal of Power Sources, 183(1), 260–274.

15. Tüber, K., Pócza, D., & Hebling, C. (2003). Visualization of water buildup in the cathode of a transparent PEM fuel cell. Journal of Power Sources, (124), 403–414.

16. Voss, H. H., Wilkinson, D. ., Pickup, P. ., Johnson, M. C., & Basura, V. I. (1995). Anode water removal: A water management and diagnostic technique for solid polymer fuel cells. Electrochimica Acta, 40(3), 321–328.

17. Rodatz, P., Büchi, F., Onder, C., & Guzzella, L. (2004). Operational aspects of a large PEFC stack under practical conditions. Journal of Power Sources, 128(2), 208–217.

18. He, W. S., Lin, G. Y., & Van Nguyen, T. (2003). Diagnostic tool to detect electrode flooding in proton-exchange-membrane fuel cells. AIChE Journal, 49(12), 3221–3228.

19. St-Pierre, J., Wilkinson, D. P., Knights, S., & Bos, M. L. (2000). Relationships between water management, contamination and lifetime degradation in PEFC. Journal of New Materials for Electrochemical Systems, 3(2), 99–106.

20. J. Larminie, A. Dicks, Fuel Cell System Explained (first ed.), John Wiley & Sons (2000)

21. Sridhar, P., Perumal, R., Rajalakshmi, N., Raja, M., & Dhathathreyan, K. S. (2001). Humidification studies on polymer electrolyte membrane fuel cell. Journal of Power Sources, 101(1), 72–78.

22. Sridhar, P., Perumal, R., Rajalakshmi, N., Raja, M., & Dhathathreyan, K. S. (2001). Humidification studies on polymer electrolyte membrane fuel cell. Journal of Power Sources, 101(1), 72–78

23. Natarajan, D., & Van Nguyen, T. (2005). Current distribution in PEM fuel cells. Part 1: Oxygen and fuel flow rate effects. AIChE Journal, 51(9), 2587–2598.

24. Owejan, J. P., Trabold, T. A., Jacobson, D. L., Arif, M., & Kandlikar, S. G. (2007). Effects of flow field and diffusion layer properties on water accumulation in a PEM fuel cell. International Journal of Hydrogen Energy, 32(17), 4489–4502.

25. Hakenjos, A., Muenter, H., Wittstadt, U., & Hebling, C. (2004). A PEM fuel cell for combined measurement of current and temperature distribution, and flow field flooding. Journal of Power Sources, 131(1–2), 213–216.

26. Jamekhorshid, A., Karimi, G., & Noshadi, I. (2011). Current distribution and cathode flooding prediction in a PEM fuel cell. Journal of the Taiwan Institute of Chemical Engineers, 42(4), 622–631.

27. Nguyen, T. V. (1996). A gas distributor design for proton-exchange-membrane fuel cells. Journal of The Electrochemical Society, 143(5), L103–L105.

28. Wood III, D. L., Yi, J. S., & Nguyen, T. V. (1998). Effect of direct liquid water injection and interdigitated flow field on the performance of proton exchange membrane fuel cells. Electrochimica Acta, 43(24), 3795–3809.

29. Liu, X., Guo, H., & Ma, C. (2006). Water flooding and two-phase flow in cathode channels of proton exchange membrane fuel cells. Journal of Power Sources, 156(2), 267–280.

30. Van Nguyen, T., & Knobbe, M. W. (2003). A liquid water management strategy for PEM fuel cell stacks. Journal of Power Sources.

31. Ashrafi, M., & Shams, M. (2017). The effects of flow-field orientation on water management in PEM fuel cells with serpentine channels. Applied Energy, 208, 1083–1096.

32. Wawdee, P., Limtrakul, S., Vatanatham, T., & Fowler, M. W. (2015). Water transport in a PEM fuel cell with slanted channel flow field plates. International Journal of Hydrogen Energy, 40(9), 3739–3748.

33. Liu, X., Guo, H., & Ma, C. (2006). Water flooding and two-phase flow in cathode channels of proton exchange membrane fuel cells. Journal of Power Sources.

34. Brennan, B. W. (1976). United States Patent [191 Stenberg [54], 146(1979), 1–6.

35. Knights, S. D., Colbow, K. M., St-Pierre, J., & Wilkinson, D. P. (2004). Aging mechanisms and lifetime of PEFC and DMFC. In Journal of Power Sources.

36. Mathias M, Roth J, Fleming J, Lehnert W. Diffusion media for PEM fuel cells. In: Vielstich W et al., editors. Handbook of fuel cells—fundamentals, technology and applications, Fuel cell technology and applications, vol. 3. New York: Wiley; 2003. [Chapter 46]

37. Qi, Z., & Kaufman, A. (2002). Improvement of water management by a microporous sublayer for PEM fuel cells. Journal of Power Sources.

38. Öztürk, A., Fıçıcılar, B., Eroğlu, İ., & Bayrakçeken Yurtcan, A. (2017). Facilitation of water management in low Pt loaded PEM fuel cell by creating hydrophobic microporous layer with PTFE, FEP and PDMS polymers: Effect of polymer and carbon amounts. International Journal of Hydrogen Energy, 42(33), 21226–21249.

39. Li, A., & Chan, S. H. (2013). Understanding the role of cathode structure and property on water management and electrochemical performance of a PEM fuel cell. International Journal of Hydrogen Energy, 38(27), 11988–11995.

40. Bae, I., Oh, K.-H., Yun, S.-H., & Kim, H. (2017). Asymmetric silica composite polymer electrolyte membrane for water management of fuel cells. Journal of Membrane Science, 542(July), 52–59.

41. Kraytsberg, A., & Ein-Eli, Y. (2006). PEM FC with improved water management. Journal of Power Sources.

42. Cha, D., Ahn, J. H., Kim, H. S., & Kim, Y. (2015). Effects of clamping force on the water transport and performance of a PEM (proton electrolyte membrane) fuel cell with relative humidity and current density. Energy, 93, 1338–1344.

512513 PEM Fuel Cell Design Exploration

Abstract:

Computer aided engineering (CAE) can be powerful tool in the design of fuel cell systems. Simulations can be used understand trade-offs between maximizing power efficiency and minimizing temperature variations when scaling from cells to stacks and modules with robust and reliable operation. Validated PEM fuel cell models are applied to 3-D stack simulations, where proprietary optimization algorithms quickly obtain better multi-parameter designs. Each simulation conserves flow, energy, species, and charge transport using segregated solvers with a high degree of linearization and efficient sparse matrix solvers. Automated workflows and state-of-the-art search algorithms allow for fast design space exploration of fuel cell dimensions, properties and operating conditions.

512873 Synthesis of Vertically Oriented Titanium Dioxide Nanostructures By Electrochemical Anodization

Abstract:

Photocatalysis has established as viable option in the development of processes for the treatment of pollutants and clean energy production. This option is based on the ability of semiconductors to generate an electron flow by means of the interaction with solar radiation. Owing to its electronic structure, TiO2 is the most frequently used semiconductors in photocatalysis, although it has a high recombination of photogenerated charges and low solar energy absorption. An alternative to reduce these limitations is the use of nanostructured morphologies which can be produced during the synthesis of TiO2 nanotubes (TNTs). Therefore, if possible to produce vertically oriented nanostructures it will be possible to generate a greater contact area with electrolyte and better charge transfer. At present, however, the development of these innovative structures still presents an important challenge for the development of competitive photoelectrochemical devices.

This research focuses on established correlations between synthesis variables and 1D nanostructure morphology which has a direct effect on the photocatalytic performance. TNTs with controlled morphology were synthesized by two-step potentiostatic anodization of titanium foil. The anodization was carried out at room temperature in an electrolyte composed of ammonium fluoride, deionized water and ethylene glycol. Consequent thermal annealing of as-prepared TNTs was conducted in the air between 450 Â°C - 550 Â°C. Morphology and crystalline phase of the TNTs were carried out by SEM, EDS and XRD analysis.

As results, the synthesis conditions were established to produce nanostructures with specific morphological characteristics.Anatase was the predominant phase of TNTs after thermal treatment. Nanotubes with 10 Î¼m in length, 40 nm in pore diameter and a surface-volume ratio of 50 are important in photoelectrochemical applications based on TiO2 due to their 1D characteristics, high surface-volume ratio, reduced radial dimensions and high oxide/electrolyte interface. Finally, this knowledge can be used to improve the photocatalytic activity of TNTs by making additional surface modifications with dopants that improve their efficiency.

512437 Techno-Economic Analysis of Renewable Transportation Fuel from Wastewater Treatment Sludge

Abstract:

For several decades, the Department of Energy Bioenergy Technologies Office (BETO) has supported research and development for the advancement of bioenergy pathways for terrestrial and algal feedstocks. More recently, BETO has begun to explore the potential of wet wastes for biofuel production, and estimates that 77 million dry tons per year of wastewater residuals, manure, food waste, and fats, oils and greases are generated annually, 65% of which are underutilized for any beneficial purpose, such as for fertilizer, biodiesel or compost. Conversion of this resource into transportation fuels could significantly contribute to the nation's renewable energy goals and provide an economically and environmentally sustainable alternative for current waste disposal practices.

Hydrothermal liquefaction (HTL) is a process that uses hot, pressurized water in the condensed phase to convert wet biomass to an oil product. As such, it is particularly well suited for processing wet waste feedstocks and eliminates the need for drying that is required for other biomass conversion technologies. The oil product, also known as "biocrude", is analogous to petroleum crude in that it contains a mixture of hydrocarbons with carbon numbers in the gasoline/jet/diesel range. However, it contains higher oxygen and nitrogen than petroleum and therefore must be hydrotreated to improve compatibility with petroleum fuels. This talk presents the conceptual process design, economics and supporting data for a sludge-to-hydrocarbon fuel blendstocks pathway through HTL and biocrude upgrading. The design and techno-economic analysis represents a goal case for the pathway, targeting performance that is anticipated to be achievable by 2022 (BETO's target year for verification of biofuel hydrocarbon pathways) with further development. The relevance of this work reaches beyond sludge to lay the groundwork for application to other distributed wet wastes and blends that together represent a substantial resource of underutilized biomass.

512913 An Integrated Techno-Economic and Life Cycle Analysis of Northeastern Willow Biomass As a Bioenergy Feedstock

Abstract:

Shrub willow biomass is being examined as a potential cellulosic bioenergy feedstock in the Northeast U.S. under a 3-year USDA Biomass Research and Development Initiative (BRDI) project that is being led by faculty at SUNY College of Environmental Science & Forestry (SUNY ESF). Integrated techno-economic analysis and life cycle assessment under uncertainty of willow biomass production for use as a bioenergy feedstock is being conducted as a task under the project. This task has determined that, while willow biomass production is not especially feasible from a financial perspective due to the presence of low fossil energy prices and uncertainty regarding future prices and willow yields, shrub willow can result in substantial net greenhouse gas sequestration when grown on abandoned pastureland or, to a lesser extent, grassland in the region. The task is also analyzing marginal carbon abatement costs for shrub willow under several different production scenarios that impact both the life cycle assessment and techno-economic analysis, including the substitution of diesel fuel by biodiesel, growth on different land types, varying fertilizer application rates, and the utilization of different shrub willow cultivars. This proposed presentation will summarize the major findings of this task under the BRDI project being led by SUNY ESF, their implications for sustainable energy in the U.S. Northeast, and the next steps under the project.

512612 Precise Determination of the Main Transition Velocities in a Bubble Column Based on New Identification Methods Applied to Ultrafast X-Ray Tomographic Data

Abstract:

Bubble column reactors are simple but very effective gas-liquid contactors. They are frequently used in the processes of absorption, oxidation, chlorination, oxychlorination, alkylation, carbonylation, hydroformylation, hydrogenation, fermentation, hydrocracking, coal liquefaction, methanol synthesis, Fischer-Tropsch synthesis, etc.

The design and control of bubble column reactors involves understanding of their complex hydrodynamic and mass transfer behavior. Different hydrodynamic regimes (homogeneous, transition and heterogeneous) and flow structures are formed in the bubble bed and they affect the degrees of mixing, mass and heat transfer. The boundaries of the different flow regimes are characterized by means of transitional gas velocities. In the literature, hitherto, only the first transition velocity has been successfully identified and an empirical correlation (Reilly et al., 1994) for its prediction has been developed. For the detection of the other critical gas velocities there are no reliable methods. That is why, in this work, new methods have been developed for the successful identification of the boundaries of the main flow regimes.

The information entropy theory was applied to time series (29,000 pixel values) measured by means of ultrafast electron beam X-ray tomography (Fischer et al., 2008; Fischer and Hampel, 2010). The X-ray data were recorded in different parts (semi-rings) of the columnâ€™s cross-section at a sampling frequency of 1000 Hz. The non-invasive X-ray tomographic scanner was located at an axial height of 0.5 m above the perforated plate gas distributor (55 holes Ã— Ã˜ 0.5 mm, open area = 0.14 %). The bubble column (0.1 m in ID) was operated with an air-deionized water system at ambient conditions. The clear liquid height was fixed at 0.66 m. The superficial gas velocities U_g were varied from 0.01 to 0.15 m/s, respectively.

The main new method is based on the reconstruction of the signal in the third semi-ring (inner radius = 10 mm and outer radius = 15 mm) and the formation of 289 state vector pairs consisting of 50 elements. The difference between the elements of every compared vector pair is fixed at 100 elements. The maximum norm with a cut-off length L equal to 5 times the average absolute deviation (AAD) is used for estimating the distance between the different state vectors. Only consecutive cases b with vector distance smaller than L are taken into account. The information entropy (IE) for each b value is calculated and then the maximum IE at each superficial gas velocity U_g is used for flow regime identification. The maximum IE exhibits three well pronounced local minima. At $U_g = 0.025$ m/s ends the gas maldistribution regime. A well pronounced local minimum at $U_g = 0.045$ m/s identifies the end of the homogeneous regime and the onset of the transition regime. The onset of the heterogeneous regime is distinguished at $U_g = 0.11$ m/s.

In the central semi-ring (radius = 5 mm) the time-dependent signal was reconstructed 5 times and then two different IE concepts (Nedeltchev et al., 2017) were applied. The absolute differences between the reconstructed data points were compared with a cut-off length L equal to 3AAD and only the values b_1 bigger than L were taken into account. In the first new IE definition, the number of repeating identical b_1 values were only considered. In the second new IE definition, all b_1 values equal to 2 or higher were taken into account. Both IE algorithms exhibit two well-pronounced local minima at $U_g = 0.04$ m/s and 0.11 m/s. They correspond to the end of the homogeneous regime and the onset of the heterogeneous regime. The correct estimation of the first transition velocity was validated by means of the overall gas holdup profile.

The above-described new methods have also been applied to the pixel data measured in all semi-rings, into which the columnâ€™s cross-section has been divided. In this way, the existence of delayed regime transitions can be examined.

Acknowledgment

The authors gratefully acknowledge the financial support of the European Research Council (ERC Starting Grant, Grant Agreement No. 307360). The financial support of the Helmholtz Association of German Research Centers within the frame of the Initiative and Networking Fund (No. ERC-0010) is also acknowledged.

References

Fischer, F., D. Hoppe, E. Schleicher, G. Mattausch, H. Flaske, R. Bartel and U. Hampel, *Meas. Sci. & Technol.* **19**, 094002 (2008)

Fischer, F. and U. Hampel, *Nucl. Eng. Des.* **240**, 2254-2259 (2010)

Nedeltchev, S., U. Hampel and M. Schubert, *Chem. Eng. Sci.* **170**, 225-233 (2017)

Reilly, I. G., D. S. Scott, T. J. W. De Bruijn and D. MacIntyre, *Can. J. Chem. Eng.* 72, 3-12 (1994)

512580 Bubbletree: A Rigorous Algorithm for Lagrangian Tracking and Statistical Analysis of Bubble or Cluster Motion and Flux within 3D Fluid Bed Simulations

Abstract:

Bubbling dynamics and distribution of gas flow critically influence the performance of fluidized bed reactors and must be quantified accurately for their design and performance optimization. In this study, BubbleTree (BT) toolset is developed using python based Visualization Tool Kit (VTK) for analyzing computational fluid dynamics simulation data and performing statistical analysis of bubbling dynamics such as their volume, location and velocity. In addition, this toolkit provides the capability to (a) track bubbles through coalescence and splitting events and (b) compute gas bypass (or throughflow) by integrating gas fluxes through the bubble surface. The BT toolset also incorporates a robust statistical method to improve the certainty of the analysis. In addition to the bubble tracking capabilities, the BT toolset can track particle clusters. Integration with the MFiX CFD simulation allows for in situ updating of physical model parameters based on bubble or particle cluster tracking. Additional developments include analyzing flow inside the bubble or cluster region and flow between regions to quantify channeling phenomena.

513483 CFD Modeling of Oxy-Natural Gas Furnace Using Detailed Kinetic Modeling with and without the Use of Chemistry Acceleration

Abstract:

Numerical modelling of the combustion in industrial geometries is often carried out using computationally less intensive methodologies such as global kinetic reaction schemes involving a few steps, or via pre-computed flamelet tables constructed from zero or one dimensional canonical systems. Although reasonable predictions of heat release, temperature, and major species like CO_2 and H_2O is possible using these combustion models, the slow forming species are often not captured accurately. Detailed kinetic modelling enables the accurate description of all the species (both slow and fast) in a reacting system but is often not used for industrial applications because of its prohibitive computational cost. The objective of this study is to investigate the effect of three chemistry acceleration techniques that can be used with the complex chemistry model in STAR-CCM+ [1] for improving CPU efficiency with potentially minimum loss of accuracy.

The test case considered in this paper is a coaxial high momentum oxy-natural gas flame in a refractory lined furnace at a thermal input of 0.8 MW [2]. Radial profiles of velocity, temperature and species are compared between experiments and simulations at several axial locations downstream of the burner. The detailed chemical kinetic mechanism GRI3.0 was used to represent the chemical kinetics and the effect of chemistry acceleration techniques such as (a) In-situ Adaptive Tabulation (ISAT), (b) Clustering and (c) Dynamic Mechanism Reduction (DMR) on the predicted results was analysed. ISAT [3] enables speed up of complex chemistry simulations by storing and retrieving approximate reaction mapping solutions related to the chemical state space during run time. The clustering method groups cells with similar thermal and chemical states before integrating the averaged state for the group which in turn causes substantial speed-up. The DMR methodology is based on Directed Relation Graph (DRG) algorithm [4] where the mechanism is dynamically reduced in every cell at every iteration by identifying and solving only those species that substantially change over the reaction step thereby reducing overall computational time. The speed up achieved for each of the acceleration techniques and the associated accuracy is reported in this study.

References

[1] https://mdx.plm.automation.siemens.com/star-ccm-plus

[2] Lallemant, N., Dugue, R., and Weber, R., (1997). Analysis of the Experimental Data Collected During the OXYFLAM-1 and OXYFLAM-2 Experiments. Technical Report F85/y/4, International Flame Research Foundation.

[3] Pope, S.B. 1997. Combust. Theory and Modelling, 1, pp. 41-63.

[4] T. Lu, C. Law, Proc. Combust. Inst. 30 (2005) 1333-1341.

511567 Computer-Aided Designing High-Performance Solid Sorbents for CO_2 Capture Application

Abstract:

CO_2 is one of the major combustion products which once released into the air can contribute to global climate change. Molecular modeling can play a role in identifying optimal sorbents. By combining thermodynamic database mining with first principles density functional theory and phonon lattice dynamics calculations, a theoretical screening methodology to identify the most promising CO_2 sorbent candidates from the vast array of possible solid materials have been proposed and validated at National Energy Technology Laboratory (NETL). The advantage of this method is that it identifies the thermodynamic properties of the CO_2 capture reaction as a function of temperature and pressure without any experimental input beyond crystallographic structural information of the solid phases involved. The calculated thermodynamic properties of different classes of solid materials versus temperature and pressure changes were further used to evaluate the equilibrium properties for the CO_2 adsorption/desorption cycles. The selected CO_2 sorbent candidates are further considered for experimental validations.

513802 Ocean Thermal Energy for Abatement of Global CO_2 Emissions

Abstract:

The worldâ€™s oceans are the largest collectors and storage of solar energy and have an enormous potential to supply growing worldwide energy demands, commodity products like ammonia, and fresh water. Global energy, environment, and economic settings have drastically changed: uncertain energy costs; increased energy demands of economic emerging countries; awareness of impact of CO_2 emission; recognition of Energy-Water-Food Nexus to support alternate energy production and growing population. There are global movements of reducing CO_2 emissions. Besides replacing coal with natural gas and CO_2 capture & sequestration (CCS) new technologies are being developed for conversion of CO_2 to value-added products. Ocean Thermal Energy Conversion (OTEC) has potentials of making significant contribution to reduce global CO_2 emission. This analysis evaluates four applications of the OTEC technology with direct impact on CO_2 abatement: a) global displacement of fossil-based power generation in the tropical island states and countries; b) at-sea production of green-ammonia using OTEC plantships displacing fossil feedstock and energy currently used for production of ammonia; c) production of desalinated water for regions of critical shortages of potable water for human consumption; and d) conversion of captured CO_2 to methanol using OTEC produced hydrogen. A case study is presented for deployment of OTEC in the Gulf of Mexico. With full implementation of the clean energy OTEC technology by 2040 and beyond in these four applications, it would play a significant role in the global CO_2 abatement.

513933 Utilization of Captured CO_2 for Manufacturing Alkyl Carbonates

Abstract:

There are very limited uses for captured CO_2 other than enhanced oil recovery (EOR) which currently has in excess of 126 projects globally. Some of these EOR projects use 2 Mega-tonne/yr of CO_2. The Global Carbon Capture and Sequestration (CCS) Institute provides a list of current and future CO_2 storage sites not linked to EOR and highlights the challenges and costs of coordinating the development of a CO_2 transportation infrastructure. Few major industrial CO_2 generation sources (e.g. fossil electric power plants, refineries, petrochemical production, ammonia/urea manufacturing facilities, etc.) are located close to these storage sites. As a result, technologies for converting captured CO_2 to value-added products are being actively pursued.

The SBIR project focuses on techno-economic merits of utilization of captured CO_2 for manufacturing alkyl carbonates. The E3Tec project team aims to convert captured CO_2 to the simplest alkyl carbonate, dimethyl carbonate (DMC) while selectively co-producing mono-ethylene glycol (MEG). An alternate process being tested uses ammonia as a chemical carrier to produce only DMC. Heat integrated reactive distillation (HIRD) equipped with side reactors and pervaporation (PerVap) membranes is ideally suited for these complex reversible and equilibrium controlled reactions.

DMC is a high-value specialty chemical with major applications in manufacturing polycarbonate, as a low VOC solvent, in lithium-ion batteries and as an intermediate in manufacturing of polyurethane. DMC is also a possible diesel fuel additive and this application would expand the DMC market tremendously. MEG is a major commodity chemical used in polyethylene terephthalate (PET) with end-use in making films, fibers, and bottles. No single chemical product will be able to make a major impact CO_2 abatement, either by direct consumption and/or off-setting CO_2 emission for the end product. However, multiple plants from distributed CO_2 source sites will constitute a significant â€œwedgeâ€ in the overall CO_2 abatement target. In order for CO_2 utilization to make a major impact, commodity chemicals such as methanol and urea would need to be the products. ; However, it would be very hard to compete against these commodity chemicals, especially with low natural gas prices. Specialty chemicals, such as DMC and MEG, with favorable product margin will provide the basis for early commercialization.

Methanol is feedstock for DMC, so once the DMC technology is demonstrated, an integrated process to make DMC can be developed with CO_2 capture linked to methanol. In essence, DMC is an ideal chemical for CO_2 utilization. The economic analysis shows that high-value products, such as alkyl carbonates, would offset the costs of CO_2 capture and sequestration.

Chromium Doped Ceria for Solar Thermochemical Fuel Production

(Principal Author)
Rahul Bhosale
Department of Chemical Engineering
College of Engineering
Qatar University, Doha, Qatar
rahul.bhosale@qu.edu.qa

Gorakshnath Takalkar, Anand Kumar, Fares AlMomani
Department of Chemical Engineering
College of Engineering
Qatar University, Doha, Qatar

Prepared for Presentation at
American Institute of Chemical Engineers
2018 Spring Meeting
14th Global Congress on Process Safety
Orlando, Florida
April 22-26, 2018

Chromium Doped Ceria for Solar Thermochemical Fuel Production

Rahul Bhosale, Gorakshnath Takalkar, Anand Kumar, Fares AlMomani
Department of Chemical Engineering
College of Engineering
Qatar University, Doha, Qatar
rahul.bhosale@qu.edu.qa

Keywords: Ceria, Chromium, CO_2 Splitting, Solar Fuels

Abstract

In this paper, the thermal reduction and CO_2 splitting ability of $Ce_{0.9}Cr_{0.1}O_{2-\delta}$ are investigated by performing multiple thermochemical cycles using a thermogravimetric analyzer. The $Ce_{0.9}Cr_{0.1}O_{2-\delta}$ material is successfully derived via co-precipitation method and analyzed via powder X-ray diffraction (PXRD). The $Ce_{0.9}Cr_{0.1}O_{2-\delta}$ derived is further tested towards its O_2 releasing and CO production capacity by performing three thermochemical CO_2 splitting cycles (regarding weight loss and gain).

1. Introduction

Production of H_2 or syngas via the metal oxide based solar thermochemical H_2O/CO_2 splitting cycle is one of the most appealing option to fulfill the future energy demand. This process is more feasible as compared to direct thermolysis of H_2O/CO_2 as the temperatures required are significantly lower. Concentrated solar power is used to drive the two-steps involved (successive reduction and re-oxidation of the metal oxides). Among the various metal oxides investigated until now[1–35], ceria-based redox materials are considered as a very good option due to their high thermal stability and faster kinetics.

To improve the fuel production ability of the ceria material, it is highly essential to enhance its thermal stability and oxygen storage capacity. According to previous studies associated with the resistive-type O_2 sensor for the exhaust gas from the automobile, selective oxidation of CO, oxidative steam reforming of methane, and others, the ceria doped with different metal cations behaves more efficiently as compared to phase pure ceria. The inclusion of transition metal cations inside the ceria fluorite crystal structure will increase the melting point and oxygen ion conductivity of the solid-solution formed. Increase in the melting point improves the thermal stability of the doped ceria material, and the rise in the oxygen ion conductivity enhances the redox reactivity via improving the oxygen storage capacity. Also, the doping of the transition metal cations can introduce the strain in the ceria host fluorite structure which leads to a more straightforward thermal reduction at lower temperatures. This can be achieved by the mismatch between the ionic radius of the ceria and transition metal cation. This mismatch introduces the oxygen vacancy in the ceria fluorite crystal structure, which can yield into an increase in the solar fuel production.

This paper deals with the synthesis of Cr doped ceria via co-precipitation of hydroxide method ($Ce_{0.9}Cr_{0.1}O_{2-\delta}$). As-synthesized $Ce_{0.9}Cr_{0.1}O_{2-\delta}$ is characterized using powder X-ray diffractometer and tested towards thermochemical CO2 splitting reaction using a high-temperature thermogravimetric analyzer.

2. Experimental

2.1.1 Materials

To synthesize $Ce_{0.9}Cr_{0.1}0_{2-\delta}$, $Ce(III)(NO_3)_2.6H_2O$ and $Cr(III)(NO_3)_3.9H_2O$ are procured from Sigma Aldrich. The precipitating agent, aqueous ammonia hydroxide (NH_4OH, 28%), is purchased from Sigma Aldrich. The gases used for the thermogravimetric analysis are acquired from Buzwair Scientific and Technical Gases supplier, Qatar. Deionized water (ultrapure Type 1 from Direct –Q system, Millipore, France) is used for the preparation of the precursor solution.

2.2. Co-precipitation synthesis of $Ce_{0.9}Cr_{0.1}0_{2-\delta}$

To synthesize $Ce_{0.9}Cr_{0.1}0_{2-\delta}$, the computed amounts of metal precursors are first dissolved in the deionized water under ordinary stirring. After achieving the complete dissolution, aqueous ammonium hydroxide is added dropwise to the metal precursor solution (pH ~ 10) until precipitation is total. The aqueous dispersion is further stirred for 48 h by maintaining the pH ~ 10. The solution obtained after 48 h is allowed to settle overnight. The precipitate obtained is washed for several times using deionized water and vacuum filtration unit. The solid particles obtained after filtration are dried, crushed using mortar and pestle into the fine powder, and then annealed up to 800 °C for 4 hours in the presence of air using a muffle furnace. The $Ce_{0.9}Cr_{0.1}0_{2-\delta}$ powder obtained after annealing is again treated by using mortar and pestle and the fine powder obtained is stored for further analysis. The variation in the color of the precipitating solution during synthesis of $Ce_{0.9}Cr_{0.1}0_{2-\delta}$.

2.3. Characterization of $Ce_{0.9}Cr_{0.1}0_{2-\delta}$

Crystallographic studies with phase composition of $Ce_{0.9}Cr_{0.1}0_{2-\delta}$ are carried out using a Panalytical XPert MPD/DY636 powder X-ray diffractometer (PXRD) with CuKα radiation (λ = 0.15418 nm, voltage = 45 kV, current = 20 mA, angular range = 20-90° 2θ, steps = 0.05° 2θ, recording time = 5 s).

2.4. CO_2 Splitting Experiments

By using the thermogravimetric analyzer (Seteram Setsys Evo.), the thermochemical CO_2 splitting ability of the synthesized $Ce_{0.9}Cr_{0.1}0_{2-\delta}$ is investigated in the temperature range of 1000 to 1400 °C. A 100 µl Al_2O_3 crucible is used to carry approximately 100 mg of $Ce_xCr_{1-x}O_{2-\delta}$ powder. This $Ce_xCr_{1-x}O_{2-\delta}$ loaded Al_2O_3 crucible is then placed on the hanging balance with very high precision. The $Ce_xCr_{1-x}O_{2-\delta}$ powder is then undergone multiple thermal reductions (in the presence of Ar) and CO_2 splitting (in the presence of a mixture of 50%CO_2 and 50%Ar) cycles in the temperature range of 1000 to 1400°C. The thermal reduction step is carried out for 60 min, and the CO_2 splitting step is performed for 30 min. The decrease in the mass during the thermal reduction step reflects into the O_2 released from the $Ce_xCr_{1-x}O_{2-\delta}$ material. In contrast, the increase in the mass during the CO_2 splitting step is attributed towards the re-oxidation of $Ce_xCr_{1-x}O_{2-\delta}$ material with CO production. To eliminate the effect of buoyancy on the experiments associated with the $Ce_xM_{1-x}O_{2-\delta}$ based thermochemical CO_2 splitting cycles, new tests are performed (using empty crucible of Al_2O_3) at operating conditions similar to that of the actual CO_2 splitting experiment.

Figure 1. Variation in the color of the precipitating solution during synthesis of $Ce_{0.9}Cr_{0.1}O_{2-\delta}$.

3. Results and Discussion

The derived $Ce_xCr_{1-x}O_{2-\delta}$ materials are characterized via PXRD analysis. The PXRD profiles for the $Ce_xCr_{1-x}O_{2-\delta}$ material are presented in Figure 3. As per the PXRD peaks reported, $Ce_xCr_{1-x}O_{2-\delta}$ material exhibit cubic fluorite structure identical to ceria. Dopant oxide peaks are absent which further confirms the formation of single-phase solid solution containing Cr incorporated in the host ceria structure. The chemical composition of the $Ce_xCr_{1-x}O_{2-\delta}$ materials is also confirmed by EDS analysis.

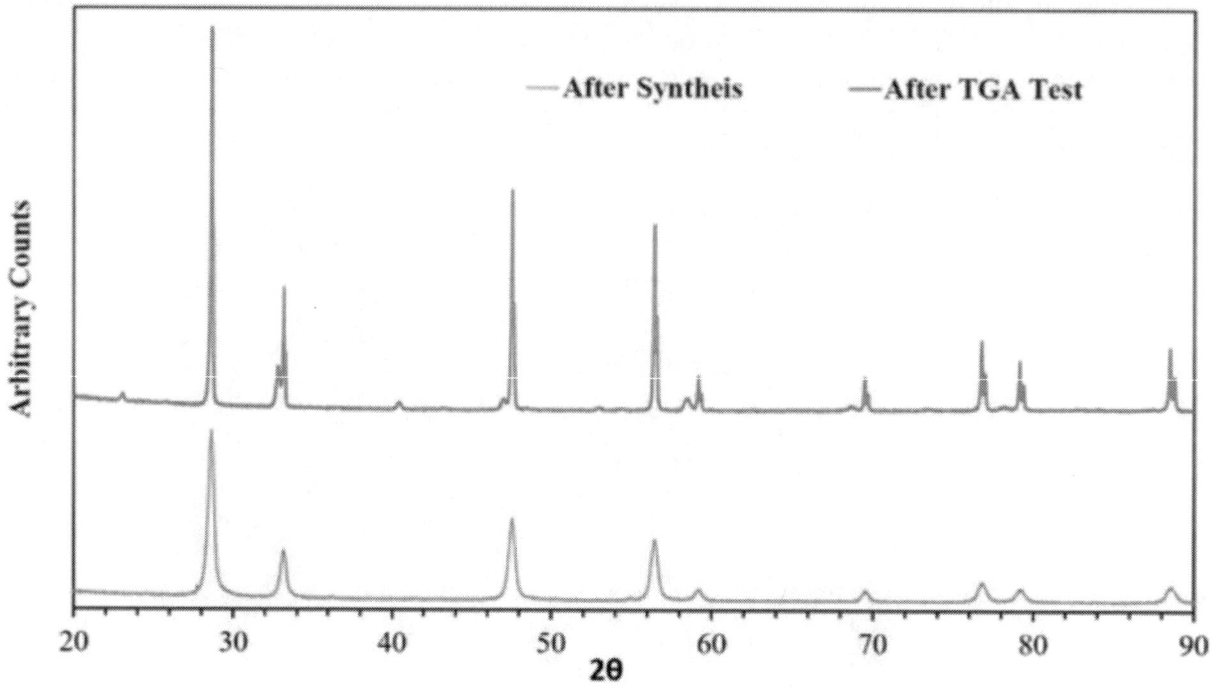

Figure 2. PXRD patterns of $Ce_{0.9}Cr_{0.1}O_{2-\delta}$, as-synthesized and after performing thermochemical CO_2 splitting cycles.

The CO_2 splitting experiments are conducted using a TGA apparatus and the TGA profile obtained during three consecutive thermochemical cycles is presented in Figure 3. During the first thermochemical cycle, 0.505% weight loss was realized during the first thermal reduction whereas 0.1% weight gain was observed during the first CO_2 splitting step. In the second step, 0.233% weight loss and 0.131% weight gain was realized during thermal reduction and CO_2 splitting phase. It is believed that the material attains thermal and redox stability and hence similar amounts of weight loss (0.144%) and gain (0.143%) was observed in the third thermochemical cycle.

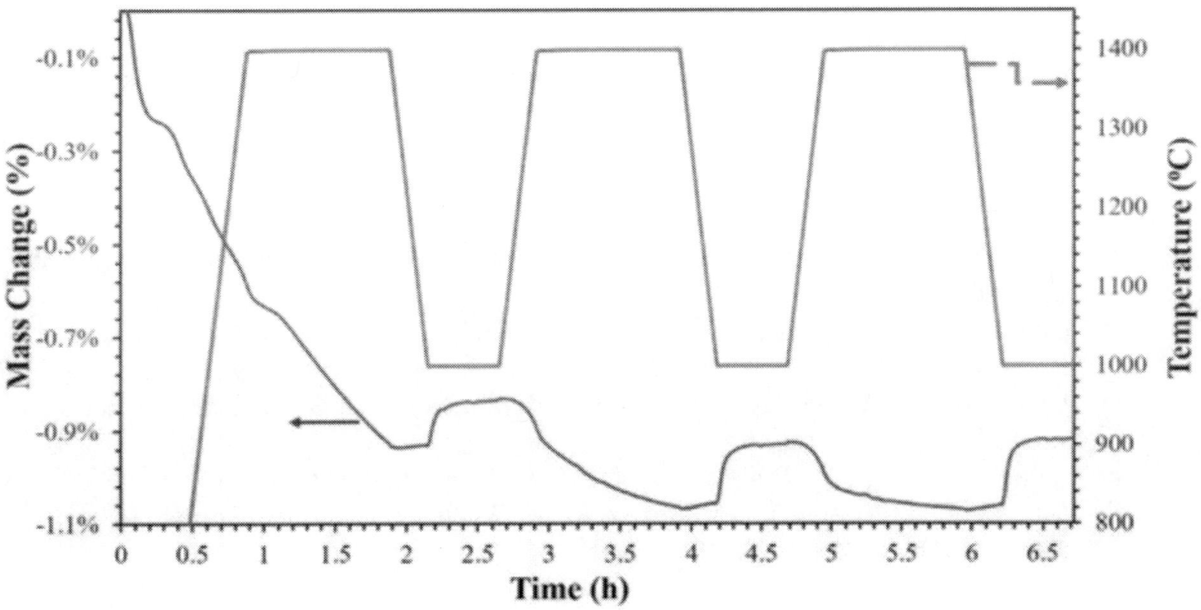

Figure 3. TGA profiles for three thermochemical CO_2 splitting cycles performed using $Ce_{0.9}Cr_{0.1}0_{2\text{-}\delta}$.

4. Conclusion

The $Ce_{0.9}Cr_{0.1}0_{2\text{-}\delta}$ with pure phase composition and nanoparticulated morphology is successfully synthesized via co-precipitation of hydroxide method. The TGA experiments indicate that the $Ce_{0.9}Cr_{0.1}0_{2\text{-}\delta}$ is capable of thermally reduced and re-oxidized in the temperature range of 1000 t0 1400 °C. Furthermore, the material seems to be thermally stability after three thermochemical cycles.

5. Acknowledgments

This publication was made possible by the NPRP grant (NPRP8-370-2-154) from the Qatar National Research Fund (a member of Qatar Foundation). The statements made herein are solely the responsibility of author(s).

References

(1) Agrafiotis, C. C.; Pagkoura, C.; Lorentzou, S.; Kostoglou, M.; Konstandopoulos, A. G. *Catal. Today* **2007**, *127* (1–4), 265–277.
(2) Kodama, T.; Nakamuro, Y.; Mizuno, T. *J. Sol. Energy Eng.* **2006**, *128* (1), 3.
(3) Gokon, N.; Murayama, H.; Umeda, J.; Hatamachi, T.; Kodama, T. *Int. J. Hydrogen Energy* **2009**, *34* (3), 1208–1217.
(4) Miller, J. E.; Allendorf, M. D.; Diver, R. B.; Evans, L. R.; Siegel, N. P.; Stuecker, J. N. *J. Mater. Sci.* **2008**, *43* (14), 4714–4728.

(5) Bhosale, R. R.; Kumar, A.; Van Den Broeke, L. J. P.; Gharbia, S.; Dardor, D.; Jilani, M.; Folady, J.; Al-Fakih, M. S.; Tarsad, M. A. *Int. J. Hydrogen Energy* **2015**, *40* (4), 1639–1650.
(6) Steinfeld, A. *Int. J. Hydrogen Energy* **2002**, *27* (6), 611–619.
(7) Abanades, S. In *Industrial and Engineering Chemistry Research*; 2012; Vol. 51, pp 741–750.
(8) Dardor, D.; Bhosale, R.; Gharbia, S.; Kumar, A.; AlMomani, F. *J. Emerg. Trends Eng. Appl. Sci.* **2015**, *6* (September 2016), 129–135.
(9) Koepf, E.; Villasmil, W.; Meier, A. *Appl. Energy* **2016**, *165*, 1004–1023.
(10) Abanades, S.; Charvin, P.; Lemont, F.; Flamant, G. *Int. J. Hydrogen Energy* **2008**, *33* (21), 6021–6030.
(11) Charvin, P.; Abanades, S.; Lemont, F.; Flamant, G. *AIChE J.* **2008**, *54* (10), 2759–2767.
(12) Bhosale, R. R.; Kumar, A.; Sutar, P. *Energy Convers. Manag.* **2017**, *135*, 226–235.
(13) Lorentzou, S.; Agrafiotis, C. C.; Konstandopoulos, A. G. *Granul. Matter* **2008**, *10* (2), 113–122.
(14) Scheffe, J. R.; Li, J.; Weimer, A. W. *Int. J. Hydrogen Energy* **2010**, *35* (8), 3333–3340.
(15) Muhich, C. L.; Aston, V. J.; Trottier, R. M.; Weimer, A. W.; Musgrave, C. B. *Chem. Mater.* **2016**, *28* (1), 214–226.
(16) Muhich, C. L.; Evanko, B. W.; Weston, K. C.; Lichty, P.; Liang, X.; Martinek, J.; Musgrave, C. B.; Weimer, A. W. *Science (80-.).* **2013**, *341* (6145), 540–542.
(17) Bhosale, R. R.; Shende, R. V.; Puszynski, J. A. *MRS Proc.* **2012**, *1387*, mrsf11-1387-e09-07.
(18) Bhosale, R. R.; Kumar, A.; AlMomani, F.; Ghosh, U.; Sutar, P.; Takalkar, G.; Ashok, A.; Alxneit, I. *Ceram. Int.* **2017**, *43* (6), 5150–5155.
(19) Bhosale, R. R.; Shende, R. V.; Puszynski, J. A. *Int. J. Hydrogen Energy* **2012**, *37* (3), 2924–2934.
(20) Bhosale, R. R.; Alxneit, I.; van den Broeke, L. L. P.; Kumar, A.; Jilani, M.; Gharbia, S. S.; Folady, J.; Dardor, D. Z. *MRS Proc.* **2014**, *1675*, mrss14-1675-rr06-10.
(21) Bhosale, R. R.; Kumar, A.; Almomani, F.; Alxneit, I. *Ceram. Int.* **2016**, *42* (2), 2431–2438.
(22) Abanades, S.; Flamant, G. *Sol. Energy* **2006**, *80* (12), 1611–1623.
(23) Ackermann, S.; Scheffe, J. R.; Steinfeld, A. *J. Phys. Chem. C* **2014**, *118* (10), 5216–5225.
(24) Bhosale, R. R.; Kumar, A.; Almomani, F.; Ghosh, U.; Al-Muhtaseb, S.; Gupta, R.; Alxneit, I. *Ceram. Int.* **2016**, *42* (8), 9354–9362.
(25) Lin, F.; Rothensteiner, M.; Alxneit, I.; van Bokhoven, J. A.; Wokaun, A. *Energy Environ. Sci.* **2016**, 2400–2409.
(26) Bhosale, R. R.; Kumar, A.; AlMomani, F.; Alxneit, I. *Ceramics International*. 2015.
(27) Ackermann, S.; Sauvin, L.; Castiglioni, R.; Rupp, J. L. M.; Scheffe, J. R.; Steinfeld, A. *J. Phys. Chem. C* **2015**, *119* (29), 16452–16461.
(28) Chueh, W. C.; Abbott, M.; Scipio, D.; Haile, S. M. *Science (80-.).* **2010**, *63* (December), 2010.
(29) Scheffe, J. R.; Weibel, D.; Steinfeld, A. In *Energy and Fuels*; 2013; Vol. 27, pp 4250–4257.
(30) Nair, M. M.; Abanades, S. *ChemistrySelect* **2016**, *1* (15), 4449–4457.
(31) Carrillo, R. J.; Scheffe, J. R. *Solar Energy*. 2017.
(32) Dey, S.; Rao, C. N. R. *ACS Energy Lett.* **2016**, *1* (1), 237–243.

(33) Dey, S.; Naidu, B. S.; Govindaraj, A.; Rao, C. N. R. *Phys. Chem. Chem. Phys.* **2014**, *17* (1), 122–125.

(34) Takacs, M.; Hoes, M.; Caduff, M.; Cooper, T.; Scheffe, J. R.; Steinfeld, A. *Acta Mater.* **2016**, *103*, 700–710.

(35) Rahul R. Bhosale, Anand Kumar, Anchu Ashok, Parag Sutar, Gorakshnath Takalkar, Majeda Khraisheh, Fares AlMomani, U. G. *MRS Adv.* **2017**, No. May, 1–8.

512859 Effect of Adsorption of CO_2 and H_2 on Methanation Reaction By Ni/Al_2O_3

Abstract:

In an attempt to study the reaction mechanism for recycling carbon dioxide to methane over Ni-Al_2O_3 catalyst, effect of adsorption of both CO_2 and H_2 was studied and their competitive uptake was investigated. Single and multicomponent adsorption experiments for both gases by 18 wt.% Ni doped onto -Al_2O_3 were conducted by using Gravimetric analysis technique.

Adsorption isotherms showed a linear coverage for H_2 onto the catalyst surface, and multilayers isotherm for CO_2 with a mono-saturation capacity of 10.2 mg CO_2/g. Multicomponent adsorption isotherm showed a total uptake of 17.8 mg/g for both gases obtained at 230 °C, total pressure of 30 bars and initial H_2/CO_2 molar ratio of 4:1. Ideal Adsorption Theory calculation (IAS) was used to predict the individual uptake and showed that increasing total pressure above 7 bars lead to decreasing the ratio of H_2/CO_2 uptake, that affected the selectivity of methane production.

Consequently, at reaction conditions of 230 °C, 7 bars, GASV of 6000 ml/g.h and H_2/CO_2 ratio of 4:1, the conversion of CO_2 was 89% with 100% selectivity to CH_4. Increasing GHSV resulted in decreasing the conversion to reach a value of 70.3% at 36,000 ml/g.h.

Synthesis of CH_4 from recycling of CO_2 is directly affected by adsorption of the reacted gases. The resistant of adsorption/desorption steps will be future investigated and compared to surface reaction one.

513638 Investigating Aerosol Formation in Flue Gas from Post Combustion CO2 Capture Plants Using Molecular Dynamics Simulation

Abstract:

Monoethanolamine (MEA) has a long history of being used as a solvent for acidic gas scrubbing. Through decades it has proven its economical and industrial viability. Nevertheless, solvent loss is observed during CO_2 capture process which becomes significant in large scale. During the absorption process, where MEA and flue gas mixture meet causing supersaturation of the gas, tiny particles named as particulate matter (PM) form as a separate phase. These particles can carry over some MEA out from the column. Few attempts have been done to predict MEA loss through simulation using commercial chemical engineering software such as Aspen Plus. However, due to the limitations of such commercial softwares a true picture of the particle formation through nucleation process is not captured. Molecular dynamics simulation can be instrumental to achieve such a goal.

In this work, molecular dynamics simulation was carried out using GROMACS software in a strive to understand the underlying mechanisms of formation of particles in CO_2 capture process. Five different systems were considered. The former three consist of only MEA and CO_2, while the latter two also contain air components, water, and sulfur dioxide. Analysis of the simulation included radial distribution function (RDF), cluster size and formation study, and energy examination. Initially, appropriate forcefield is selected by computing densities and self-diffusion coefficients of MEA and CO_2 and comparing them with their known value. As a result, literature data has been chosen, which yielded the following data for MEA and CO_2 density and self-diffusion coefficient: 1094.50 ± 36.27 kg/m^3 and $(0.261 \pm 0.005)\times10^{-5}$ cm^2/s, 1.624 ± 0.10 kg/m^3 and $15550.96 \pm 436.89\times10^{-5}$ cm^2/s, respectively. Graphical representation illustrates formation of molecule clusters (PM) in all the systems. These clusters contained all the components, but were mainly consisted of CO_2, water and MEA molecules. Furthermore, existing particles in the flue gas, which are represented by SO_2 particles are also engulfed into the clusters. RDF analysis demonstrated distinct peaks between 4 and 5 Ã... *for all systems in both MEA...CO_2 and CO_2...CO_2 correlations. In cluster size analysis of MD ensemble, number of clusters decreased suddenly indicating quick formation of clusters within 2 ns. Notably, in case of presence of air components and water, cluster sizes were more stable. Energy analysis revealed that MEA â€“ CO_2 interaction is suppressed when air components are introduced to the system. It increased from -76.7094 kJ/mol to -15.0158 kJ/mol, which means inhibition of attraction between MEA and CO_2. CO_2 â€“ CO_2 potential dropped from -23851.0607 kJ/mol to -12113.3779 kJ/mol. Most likely, this is because CO_2 molecules are also attracted to the other components such as nitrogen and oxygen.*

AUTHOR INDEX

AIChE
120 Wall Street, FL 23
New York, NY 10005-4020

CURRAN ASSOCIATES INC.
proceedings.com

ISBN 978-1-5108-6431-3

9 781510 864313